AF574335

Lubricating Grease Guide

All Rights Reserved. No part of this publication may be reproduced without written permission from the National Lubricating Grease Institute.

First Edition, 1984
Revised 1987
second printing 1989

Library of Congress Catalog Card Number: 84-61641

Published in Kansas City, Missouri, U.S.A.

Copyright 1984, 1987, 1989, National Lubricating Grease Institute

ISBN 0-9613935-1-3

Copies available from:

NATIONAL LUBRICATING GREASE INSTITUTE
4635 Wyandotte Street, Kansas City, Missouri 64112

FOREWORD

The National Lubricating Grease Institute was founded with the objective to foster cooperation in the exchange of technical information among its members. This guide is evidence that the Institute's objective, established over fifty years ago through the wisdom of our predecessors, has not changed. One has only to review the list of contributors to appreciate the effort and cooperation involved in developing this guide. It was conceived by the Board of Directors who assigned the various chapters to its members. The Steering Committee of the Technical Committee then provided technical and editorial review. The various chapters were then organized and the guide partially rewritten to its present unified form by our Editor, Mel Ehrlich. It was then my pleasure to act as Chairman of the Committee for the Board of Directors to complete the final draft for publication. This effort proceeded rapidly through the cooperative efforts of Jack W. Lane and Richard B. Batts with our Editor, Mel Ehrlich, and our first Committee Chairman, George R. Berger.

The result shows that the Institute drew on the strength and expertise of its membership to cover the subject. People entering the field will find it to be a valuable source of material to get acquainted with essentially all aspects of lubricating grease.

Publication of this second edition has become necessary as a result of the popularity of the first edition. This updated and revised version of the *Lubricating Grease Guide* has been improved with the addition of an index, which we believe will greatly increase its utility. We wish to thank those who participated in the preparation of this revised edition.

W. Pat Scott
Past Chairman, NLGI Technical Committee

Jon C. Root
Chairman, NLGI Technical Committee

PREFACE

The purpose of this guide is twofold: to introduce the reader to a broad spectrum of subjects related to grease and to serve as a quick reference.

It has been my experience that many students of lubrication grasp the principles of fluid lubrication quite easily, but find grease manufacture, formulation and application somewhat of a mystery. Because of this confusion, the National Lubricating Grease Institute (NLGI) presents an intensive, well attended, grease education course in conjunction with its annual convention.

This guide is not intended to be a substitute for that course. Rather, it is an adjunct to the course. It will serve as a reference by all manufacturers, marketers and users in the conduct of their daily business and serve as an educational source for the less technical student.

The chapters have been arranged in a sequence that will lead the reader through the subject of grease in a logical manner. Each chapter, however, stands on its own so that the guide can be used for quick reference.

I thank the experts who contributed their expertise, time and energy to each chapter and their companies that supported them. I also thank Burr J. French of Ethyl Corporation for his assistance in technical writing, Jack W. Lane, C.R.D. TOTAL FRANCE for his editing assistance and Mel Ehrlich who is respon-sible for the final technical writing and editing.

George R. Berger (Retired)
Sun Refining & Marketing Co.

ACKNOWLEDGEMENTS

The Officers and Board of Directors of the NLGI have long perceived a need for this publication. The books broadly available are reference materials, quite useful to those actively working with greases. But people entering the grease industry have had to be trained by individuals and companies on a person-to-person basis or in small classes. This guide has been needed to help people get started, and then for use as reference material.

As this objective came into focus, means of accomplishing it were conceived. Some Board members wrote material, others arranged for experts in their companies to provide text, diagrams, tables and photographs. The material was then organized into this publication.

The support and enthusiasm of the Officers and Board over the time required to accomplish all this is gratefully acknowledged. Those companies and experts who are known to have contributed are listed below. We hope they are all here. For any omissions we can only apologize.

Melville Ehrlich
Editor

Contributors

G. R. Arbocus, Chief Engineer, Keystone Div., Pennwalt Corp.
P. A. Bessette, Laboratory Director, William F. Nye, Inc.
W. T. Brannen, Executive Vice-President, The Elco Corporation
P. A. Cook, Technologist, Texaco Inc.
B. J. French, Technical Analyst Writer, Ethyl Corp.
R. M. Hooks, Lubricants Chemist, NCH Corporation
E. R. Howell*, President, Jesco Resources Inc.
M. A. Julich*, Chief Chemist, Cato Oil & Grease Co.
A. Lewis, E/M Lubricants, Inc.
K. A. MacKenzie, Research Associate, Shell Canada Ltd.

T. G. Musilli, Vice President of Technical Services, Battenfeld-American, Inc.
E. H. Myers, Technical Director, Brooks Technology Company
R. B. Portwood, Sr. Vice President Manufacturing & Technical Director, Pacer Lubricants, Inc.
J.C. Root, Director of Technology, Southwest Div., Witco Corp.
F. S. Sayles*, Mobil Oil Corp.
R. T. Schlobohm, Shell Development Company
H. H. Schreimann, President, MSI Ltd.
W. P. Scott, Senior Research Scientist, Conoco Inc.
C. M. Solzman, Senior Specialist Technical Lubricants, Chevron U.S.A. Inc.
J. B. Stucker, Union Oil Company of California
L. E. Tedrow, Product Engineer, Mobil Oil Corp.
H. R. Titsworth, Amoco Oil Company
P. F. Vartanian, Chevron Research Company
T. M. Verdura, Senior Research Scientist, General Motors Research Laboratories
R. E. Wagar, Product Engineer, Mobil Oil Corp.
T. C. Wilson, Chief Chemist, Pittsburgh Manufacturing Plant, Exxon Company, U.S.A.

Sponsors

R. B. Batts, Vice President and General Manager, Brooks Technology Company
J. A. Bellanti, President & Chief Operating Officer, Battenfeld Grease & Oil Corporation of New York
R.W. Johnston*, Vice President, Southwest Div., Witco Corp.
G. S. Kent, Manager, Product Commercialization Technical Service, Chevron U.S.A. Inc.
R. E. Kuhlman, North American Sales Manager, Ethyl Petroleum Additives
J. W. Lane, C.R.D. TOTAL FRANCE
W. A. Magie, President, Magie Brothers Oil Company
C. W. Martin, Operations Manager, Universal Cooperatives, Inc.
R. V. Merrell, Director of Marketing & Sales, Ultrachem Inc.
C. P. Monti, Manager, Product Dev. & Control, Texaco U.S.A., a Div. of Texaco Inc.
R. L. Waring*, Vice President Engineering & Development, Cato Oil & Grease Co.

Please note some of the above contributors were also sponsors.
*deceased

CONTENTS

Chapter 2 - Manufacturing Modern Greases

Chapter 3 - Testing Lubricating Greases

Chapter 4 - Characteristics of Today's Greases

Chapter 5 - When To Lubricate With Grease

Chapter 6 - Which Grease To Use When

Chapter 7 - Troubleshooting Guide

Chapter 8 - Packaging, Storing, Handling, and Dispensing Grease

Shown on the front cover is an electron microscopic view of the thickener structure of a lithium hydroxystearate lubricating grease. Artwork courtesy of Conoco Inc.

The information in this publication was drawn from sources believed to be reliable. However different manufacturers use a variety of formulae and process conditions in making their products. There can be no assurance that any one grease found in the field will show performance or test results precisely as given here. These results should be considered as guides, not absolutes.

CHAPTER 1

BACKGROUND

HISTORICAL

Man's need to reduce friction dates back to ancient times. Pictures on the inner walls of tombs tell us that as early as 1650 B.C., attempts were made to reduce friction. As far back as 1400 B.C., both mutton fat and beef fat (tallow) were used in attempting to reduce axle friction in chariots.

Analyses of residues from ancient axle hubs indicate that fat and lime were sometimes mixed together to make a more complex lubricant. However, there is no indication that technology then existed which would allow the manufacture of a modern type of lubricating grease.

Until the mid-1800s, the universal lubricants continued to be mutton and beef fats, with certain types of vegetable oils playing minor roles. In 1859, however, Colonel Drake drilled his first oil well, and the face of the world began to change. Since that time, most lubricants have been based on petroleum ("mineral") oil.

The word "grease" is derived from the early Latin word "crassus," meaning fat. Throughout this publication, the designation "grease" refers only to lubricating grease.

Lubricating grease was defined in ASTM* D288 as "a solid-to-semifluid product of dispersion of a thickening agent in a liquid lubricant. Other ingredients imparting special properties may be included." Such definitions no longer appear in the book of ASTM standards.

*American Society for Testing and Materials, 1916 Race Street, Philadelphia, PA 19103

This definition is technically accurate and, to an expert in the field, sufficient. However, most of us could use some amplification and clarification of these words.

The material we disperse in a liquid lubricant is usually a solid. The solid must be available in small particles so that it can form a stable dispersion--one which will not settle out on standing. In order to develop thickening, the solid and the lubricating liquid had best have some affinity for each other. That affinity also helps keep the dispersion stable.

Affinity of the liquid for thickener is often called its solvency; some oils are better solvents than others. Affinity of the solid for the liquid is often called its solubility. At acceptable concentrations, solids which are either completely soluble or completely insoluble in a given liquid are not usually satisfactory thickeners for that liquid. To achieve a "solid-to-semifluid" product, affinity between solid and liquid must be moderate.

Since the dispersion formed is to be a lubricating grease, the product must be able to lubricate. It is possible to make a greaselike material by dispersing coal dust and/or rock dust in a mineral oil. The product, sometimes formed in coal-mining equipment, looks like a grease, feels like a grease, and even seems to meet the definition of a lubricating grease, but fails in that it is not a satisfactory lubricant, leading to high wear.

Mineral oils thickened with calcium soap (commonly known as lime soap) constituted the first type of lubricating grease produced and marketed in any volume. Since they were the only such products available for many years, they have enjoyed the largest sale of any one type of soap-thickened oil. They continue to be relatively popular, although far outshadowed now by newer products.

The first calcium soap greases resulted from cold mixing. The soap could have been formed from rosin oil (an acid) in solution in mineral oil, reacting with lime, which would have been dispersed in water and/or oil. When mixed, the two components would "set" to a firm grease. The mixing was done by hand in the early days, later by machine. No external heat was required, except for dissolving the rosin oil.

The product of that reaction would not be considered a great lubricant in today's picture. But for the slow-moving machines of the 1800s and early 1900s, it was adequate. It is still made to a limited extent, often being designated "axle" grease. Any water used and water developed from the reaction remain in the product. If the application is hot enough to evaporate that water, the grease disintegrates.

As automobiles, railroad engines, and industrial machinery evolved, operating at higher speeds and temperatures, better greases were needed. Calcium soaps made from animal fats in a cooking process replaced much of the cold mix product. These greases are smooth, effective, water-resistant lubricants. Unfortunately, the dispersion of such calcium soaps in oil must be stabilized, usually with water. The amount of water is small, in the order of 1 to 2 percent. Nevertheless, this grease cannot be used in applications which would raise its temperature near the boiling point of water. Calcium greases can hold their constituent water for a long time in service, as long as temperature is moderate.

To get away from that temperature limitation, anhydrous greases were sought. Two types were found, each with its own advantages and weaknesses.

Aluminum stearate soap makes a clear, smooth, attractive grease. It is very water-resistant and fairly rust-resistant. Unfortunately, its heat resistance is not much better than was found with lime soap greases. In this case, there is no water to be evaporated, but as the grease is heated above about 175°F (79°C), it tends to get rubbery, requiring higher power for machinery in which it is the lubricant. On cooling, this rubbery product becomes brittle, cracks up, bleeds out a significant proportion of its oil, and develops a grainy, unattractive structure.

Aluminum stearate thus never attained a large share of the lubricating grease market. It is still made, though, and will probably continue to have its supporters wherever its strengths fit the needs of the application. Its production volume is low and falling, although it may have leveled off.

Aluminum complex grease is a different type of product and will be discussed later.

Sodium soap greases constitute an earlier anhydrous class, developed in an attempt to replace lime soap greases. Instead, sodium grease found its own niche. Sodium (or soda) soaps are soluble in water; their greases, not surprisingly, are sensitive to water. Some emulsify and wash away even in room-temperature water, others only at higher temperatures.

Compensating for this weakness, they could be used under dry conditions at temperatures some 100°F (56°C) above calcium greases. Sodium greases are often more stable to mechanical handling, or working, than calcium or aluminum greases.

Sodium soap greases lubricated automotive wheel bearings, electric motors, and much industrial machinery for many years. They are being replaced by greases which are not water-washable and have equal or better heat resistance. Lubricating greases with this combination of characteristics are called "multipurpose."

In the 1930s and 1940s, researchers started to find new thickeners for multipurpose greases. The first may have been calcium complex greases. Very soon after came lithium and barium greases. The exact chronology here is not important. What is significant is the rapid and almost concurrent discovery of new thickeners, which were available for commercial exploitation at the end of World War II. In 1940, the first calcium complex grease patent was issued. Also in 1940, C. E. Earle, a research engineer with the U.S. Naval Testing Station, applied for patents on lithium soap greases. The patents were not issued until 1942 and 1943. Certain lithium greases proved very helpful during the war period, primarily for use in aircraft lubrication.

One need only look at a recent NLGI production survey to realize the significance of Mr. Earle's work. Lithium greases constitute well over 50 percent of total production. To say this differently, more lithium grease is manufactured than all other types together. These products combine heat-resistant structures such as are found in sodium soap greases with the water resistance of calcium soap greases.

Most sodium and calcium greases are made from animal fats and/or fatty acids. Sometimes, hydrogenated fatty acids resembling stearic acid are blended in. Shortly after lithium soap greases were developed, it was found that hydrogenated castor oil or its fatty acid made unusual and useful lithium soap greases. Most lithium soap greases are now made from this type of soap, known chemically as lithium 12-hydroxystearate.

Barium greases never achieved the popularity of lithium greases. They are heat-resistant and highly water-resistant but have some inherent low-temperature problems, which may be due, at least in part, to the high molecular weight of barium soap. Most barium greases are complexes rather than simple soap greases.

Other simple soap thickeners were discovered in the 1930 to 1950 period; they were produced in only limited volume and have not had significant impact. Other complex thickeners, however, have achieved a measure of popularity.

Calcium complex grease became available in commercial quantities in the 1940s. More suppliers entered the market in the 1950s. In recent years, however, this grease's share of the market has been falling. This type of grease is both water- and heat-resistant. Nevertheless, the product is sensitive to water, becoming firmer or softer under different conditions, and, if overheated, tends to harden. Calcium complex greases carry high loads in operating bearings. They are thus designated EP (extreme pressure) products.

Aluminum complex greases were patented in 1952. Volume is significant, although not a large percentage of total grease production. Volume appears to be rising. These products have outstanding water resistance, particularly in withstanding water sprays in rolling mills.

A more recent development is lithium complex grease, which first came on the market in 1962. Such grease has the usual properties of lithium greases, with added resistance to heat.

Other complexes have been discussed but have not become large volume products.

Non-soap thickeners, too, have become significant factors in the lubricating grease market. Some are pigments, others miscellaneous chemicals, such as terephthalamates. Although they fulfill a real need, most of them have been made and supplied in only small volumes. Two of the non-soap thickeners (the name tells what they are not rather than what they are) are significant: modified clays and substituted ureas.

The modified clays are often designated as organophilic bentonite, montmorillonite, etc. The clay, in its natural state, is sensitive to water. To avoid this water sensitivity, the clay is reacted, coated, or otherwise treated with a suitable quaternary ammonium compound, changing its nature from hydrophilic (water-loving) to hydrophobic (water-hating). Such a treated clay is very resistant to heat and adequately resistant to water.

The earlier substituted ureas were often aryl ureas; newer ones are alkyl aryl polyureas. These thickeners are chemicals, not soaps, and are free of metallic content. Possibly for this reason, they are highly oxidation-resistant. However, the formulation actually marketed may contain metallic compounds among its additives. These greases are particularly recommended for the lubrication of ball bearings, such as are found in electric motors.

As a modification of polyurea, we find a complex consisting of polyurea thickener and calcium acetate. This complex has EP properties such as we find in calcium complex greases.

In the next several chapters, these greases will be discussed in more detail.

TECHNICAL

The formulation, manufacture, and behavior of lubricating greases involve a significant amount of technology, grounded in various branches of physics, chemistry, and chemical engineering.

We will look at the raw materials used in this industry and the reactions they undergo in forming thickeners.

Petroleum Oils

A typical grease might contain 90 percent petroleum or "mineral" oil. Such oils are mixtures of hydrocarbons, which, as the name indicates, consist of **hydrogen** and **carbon.** Carbon is capable of forming chains which combine with other elements, such as hydrogen and oxygen. The length of the chain is important in forming a lubricant. Consider several typical chains.

If one atom of carbon holds all the hydrogen it is capable of holding, methane (CH_4) is formed. This is a combustible gas, a constituent of natural gas. If the carbon should form a chain of six carbons which contain all the hydrogen they can hold (saturated), we would find hexane (C_6H_{14}), a flammable liquid which might be a constituent of gasoline. If the carbon chain should be 14 carbons long, we would find $C_{14}H_{30}$, a possible constituent of kerosene. A chain of 25 carbons would give $C_{25}H_{52}$, found in lubricating oil. Chains of greater length form oils of higher viscosity. With still longer chains, we would have wax.

There are many modifications of such hydrocarbon chains. Chains may be straight or branched; carbon atoms may form a ring; and the amount of hydrogen attached to the carbon backbone may be all the hydrogen that can be retained by the carbon (saturated) or less than the full complement of hydrogen (unsaturated). Of these, the straight-chain saturated hydrocarbons are the poorest solvents. Branched-chain saturated hydrocarbons are almost as poor solvents. All these saturated hydrocarbons are known as paraffins. An oil consisting mainly of paraffinic hydrocarbons is called a paraffinic oil (Figure 1.1).

If an oil lacks some of the hydrogen its carbon chain could hold, it is called unsaturated or olefinic. Olefinic oils are better solvents than paraffinic oils, but poor in oxidation stability.

The carbon chains can form rings of different types. The cyclic hydrocarbons may be naphthenic or aromatic. Both are better solvents than paraffinic hydrocarbons, but they vary widely in degree of solvency and the kinds of chemicals for which they are good solvents. Oils with high proportions of ring hydrocarbons, called naphthenic oils, may include both naphthenic and aromatic hydrocarbons. Paraffinic oils are never exclusively paraffinic, containing sizable proportions of naphthenic hydrocarbons.

Synthetic Lubricating Oils

Most lubricating greases are made from petroleum oils. However, for some applications, better performance in one or more areas is

Figure 1.1 - Representative Hydrocarbon Structures

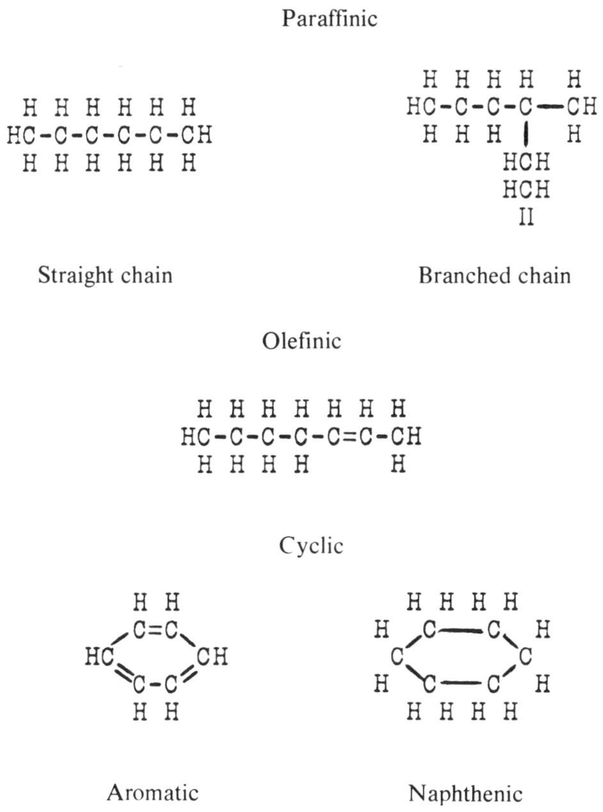

Paraffinic

```
 H H H H H H
HC-C-C-C-C-CH
 H H H H H H
```

Straight chain

```
 H H H H  H
HC-C-C-C—CH
 H H H |  H
      HCH
      HCH
       H
```

Branched chain

Olefinic

```
 H H H H H H H
HC-C-C-C-C=C-CH
 H H H H     H
```

Cyclic

```
     H H
    ,C=C,
 HC'     `CH
   `C-C'
     H H
```

Aromatic

```
   H H H H
 H  ,C—C,  H
   C'    `C
 H  `C—C'  H
   H H H H
```

Naphthenic

required, justifying the use of a more costly lubricating fluid. Thus a variety of chemicals has become available, each unique, but all categorized as synthetic oils. Greases have been made from all of these oils, the thickeners being selected, in most cases, from the ones commonly available. However, a few of these synthetic oils are thickened with specific and unusual thickeners. These costly, low-volume products will not be discussed here. They include fluorinated hydrocarbons and ethers, highly resistant to oxidation, and polyphenyl ethers, which are radiation-resistant.

Among the common synthetic oils worthy of mention are polyglycols, both water- and oil-soluble; organic esters, both polyol esters and dibasic acid esters; synthetic hydrocarbons (SHC); and silicones. All of them have higher useful temperature ranges than petroleum oil. The synthetic oils have relatively low evaporation rates and high oxidation stability at high temperatures while maintaining low viscosity at low temperatures. Each of these families has its own unique characteristics.

Polyglycols are formed by polymerizing ethylene glycol and/or propylene glycol into water-soluble and water-insoluble fluids. Under high-temperature oxidizing conditions, they decompose into volatile compounds, leaving no residue. Polyglycol greases, of course, may leave some residue from the thickener used. They are stable at fairly high temperatures and have little tendency to swell many elastomeric seals. They are sensitive to water and may soften paints, varnishes, and similar finishes.

Organic esters are formed from dibasic acids with monohydroxy alcohols or from monobasic acids with polyhydroxy alcohols. As examples, the former could be made from azelaic acid and 2-ethylhexyl alcohol, the latter from pelargonic acid and pentaerythritol. Although they differ in some significant properties, both are subject to hydrolysis, affect paint and finishes, and cause seal swelling. They are used in low-temperature or wide-temperature-range greases, particularly for use in aircraft.

Synthetic hydrocarbons are available in several structures. Two have been used for greases: polyalphaolefins and alkylated aromatics. Both are hydrocarbons and show little effect on paints and finishes or on seals. They are hydrolytically stable--not split by water. There are some significant differences, however.

Polyalphaolefins are made through limited polymerization of an alphaolefin. For example, octene-1 might be tetramerized, producing a hydrocarbon molecule of 32 carbons. Or decene-1 might be trimerized, producing a molecule of 30 carbons. These would be similar in viscosity, differing in some properties such as evaporation and flash point. These oils are usable over a wider temperature range than mineral oils; they are less volatile, more heat-resistant and oxidation-stable at high temperatures, and more fluid at low temperatures. They can be thickened with a variety of thickeners, forming greases which are used on planes and ships and are finding many industrial applications.

Alkylated aromatics, such as dialkyl benzene, result from quite different chemistry from the oils described above. Their properties are similar, but viscosity indices are lower, resulting in somewhat more viscosity change with temperature.

Silicones do not have a chain of carbon atoms, but rather a chain of alternating silicon and oxygen atoms. To this extent, silicone fluids are polymers. They have excellent fluidity at low temperatures and low volatility and good oxidation and thermal stability at high temperatures. Thus they are useful as the fluid component for high-temperature and wide-temperature-range greases. They are quite water-resistant and have little effect on many common elastomers.

Silicone fluids and the resultant greases are not generally useful where loads are high. In such circumstances, wear with silicones is also high. These materials do not respond well to most load-carrying additives which help petroleum oils and greases. Another strange deficiency is found when silicones are used as lubricants. If some of the lubricant gets on a metal surface, paint usually will not adhere to that surface. However, a silicone lubricant will not affect already-painted surfaces, except if someone tries to repaint them. Silicone fluids are quite costly.

Acids

Most greases are made from petroleum oils thickened with soaps. In turn, these require preconsideration of fats, fatty acids, and neutralization reactions. Fatty acids may be a good place to begin, but we really should go back one more step to consider acids in general.

Every acid contains hydrogen. In solution, this hydrogen takes a charged form we call a hydrogen ion, $H+$. Most of the common acids are inorganic. They vary widely in composition and in strength.

One group, the organic acids, contains a unique configuration of atoms, the COOH group. The carbon of this group can be attached to a chain of carbons or a ring of carbons or can be attached to an atom other than carbon. In the latter case, we could have formic acid, HCOOH, a water-soluble, corrosive, hazardous chemical.

If the chain has two carbons, we have acetic acid, CH_3COOH, water-soluble, not really poisonous,but quite hazardous. Concentrated, it is unpleasant to breathe, corrosive to many metals, and hazardous to the eyes, skin, nasal passages, mouth, etc.

If the carbon or hydrocarbon chains are long, acids take on a different character. An important acid in grease formulation is in a chain of 18 carbons--stearic acid, $C_{17}H_{35}COOH$. It is a weak acid, not corrosive to the skin or most metals. It has little odor, and that mild odor resembles fat. In fact, it may be made from animal fat. It feels slippery or fatty. We know such long-chain organic acids as fatty acids.

One modification of stearic acid which is important in the grease industry is 12-hydroxystearic acid. This material, derived from castor oil, is oil-soluble, not water-soluble; has a mild, though somewhat irritating, odor; and is not harmful to the skin. In the molecule of stearic acid, a hydroxyl (OH) group is present in the position identified as No. 12.

Another organic acid of importance is benzoic acid. This consists of a COOH group attached to an aromatic ring of six carbon atoms (benzene) rather than a chain.

Bases or Alkalies--Neutralization

The opposite of an acid is a base or an alkali--a strong base. A base is a combination of a metallic element or radical with one or more hydroxyl (OH) groups. Bases of interest in making greases include calcium, sodium, and lithium hydroxide.

When an acid and a base are brought together in equivalent quantities, the acid and base are neutralized. The hydrogen ion and the hydroxyl ion merge to form water. The rest of the acid molecule and the balance of the base or alkali molecule also merge to form a salt. This can be illustrated as follows.

HCl	+	NaOH	⟶	NaCl	+	HOH
Hydrochloric acid	+	Sodium hydroxide	⟶	Sodium chloride	+	Water
an acid	+	a base	⟶	a salt	+	(also written H_2O)

Sodium chloride, better known as table salt, is a representative of an enormous family, all of which are called salts. They are defined as the products, other than water, of the neutralization of an acid and a base.

When the acid used is a fatty acid, the "salt" formed is not called a salt, but rather a soap. Because of their long hydrocarbon chains, soaps are relatively soluble in oils. Salts are apt to be insoluble in oils. The chemical structure of one soap, lithium 12-hydroxystearate, is given in Figure 1.2.

Figure 1.2 - Chemical Structure of a Soap

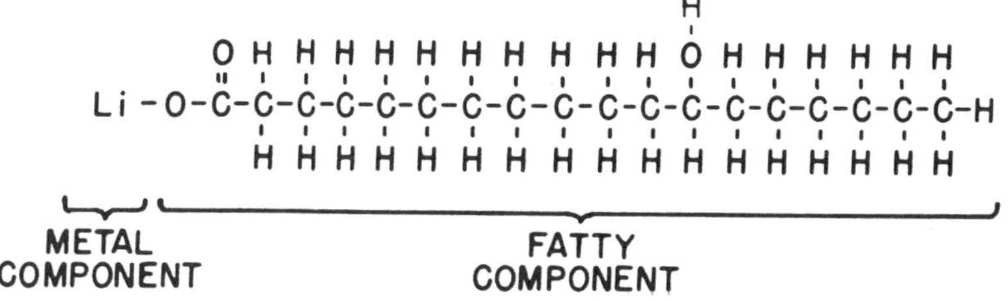

LITHIUM 12-HYDROXYSTEARATE SOAP

Fats, Fatty Acids, and Soaps--Saponification

In nature, fatty acids are common, but not plentiful. Instead, we find a wide variety of fats and fatty oils. Solid fats include beef tallow, lard, and butter. Liquid fats, or fatty oils, would include olive, castor, soybean, and peanut oils. All of the liquid or solid fats are glycerides--combinations of the fatty acid and glycerin.

If each of these fats is cooked with water and perhaps a catalyst, the fatty acid and glycerin are split apart. The two materials are separated; the fatty acid can then be neutralized to form a soap, and the glycerin will not be present.

Alternatively, the fat can be cooked with an alkali. Then the corresponding soap and glycerin are formed; the fatty acid step has been eliminated. This process of forming soap by splitting a fat is known as saponification. The glycerin remains in the soap-oil mixture. In making greases, we sometimes use fats, sometimes fatty acids, and at other times, mixtures of the two.

Soaps as Thickeners

If the saponification or neutralization reaction just described is carried out and the soap is then mixed with oil, we are not likely to obtain a satisfactory grease structure. There are two essential reasons: The soap must be in extremely fine particles, and there must be a certain degree of mutual attraction between thickener and fluid. These forces of attraction have been described as solvency of the fluid and as solubility of the thickener.

Such forces require balancing. If solubility of the soap in the fluid is excessive, thickening will not occur. If a less soluble soap is used or a liquid of lesser solvency, thickening may be accomplished. If we go too far and the soap used is not soluble enough in the fluid used, we again do not obtain thickening.

Attractive forces between a solid and a liquid are effective only at the surfaces of the solid. If the forces are to be strong, a large area of solid surface must be exposed to the liquid. If a soap is crystallized from solution in oil, the crystals formed are apt to be long, irregular, twisted, or tangled fibers. A large number of fine fibers add up to an enormous surface area. The total force exerted over such a large area of interface between solid and liquid is astonishingly large.

Other factors come into play. Twisted fibers can become tangled, mechanically trapping oil. And fibers may have chemically attractive points along their length, further causing fibers to form a three-dimensional network within which oil is held strongly.

When grease ingredients are reacted hot in the presence of oil, the soap crystallizes from oil solution as described above--in fibers of characteristic length, diameter, configuration, and interaction. The chemistry of each of these systems, along with this physical form, determines the properties of the grease. Temperature of processing, cooling rate, and intensity of mixing or milling all affect size and shape of fibers.

Some fibers are relatively weak and can be broken up or chopped up easily. Greases made from such fibers will have poor mechanical stability. If the thickener in a grease crystallizes with large fibers, milling may improve fiber size, forming a sizable number of smaller fibers. After milling, surface area will be greater, leading to a larger total of attractive forces. Then milling will lead to firmer grease. As milling continues, soap crystals are apt to be chopped down to where tangling decreases. Then further milling will lead to more oil separation and a weaker structure.

Milling and Homogenizing

In grease work, milling and homogenizing converge to where they are virtually indistinguishable. A mill is a device in which large solid particles are made much smaller. In grease manufacture, a dispersion or slurry of thickener in fluid is passed through a narrow space in which a high-speed rotor turns against a stator. Intense shear in this small space not only breaks down the solid particles or fibers, but disperses the resultant small particles in the liquid.

Forming dispersions and emulsions is usually carried out in a homogenizer. Thus grease mills often serve as both mills and homogenizers. There is not much merit in trying to determine what proportion of the action in a given device is milling and what is homogenizing. Both mills and homogenizers are used in grease manufacture.

Inorganic Thickeners

Thickener particles are not all fibrous. Some thickeners consist of tiny spheres; others are platelets. As long as particle size is small enough, the large forces built up along the great area can hold fluid closely, forming a grease structure. One remarkable example is silica. In its usual form, it is sand--hard, abrasive, and incapable of being dispersed in petroleum oil. But through some complex and costly chemical processes, this material is converted to a form resembling an aerogel--a light, fluffy, noncrystalline powder. In this form, surface area is almost unbelievably large, forces are intense, and an effective grease structure is developed, even though silica is not at all soluble in oil.

Another surprising grease thickener is made from certain clays, such as montmorillonite. Their inherent structure is in fine plates or platelets, which attract water (hydrophilic) but not oil. The clay can be dispersed in water, forming a gel or, if dilute, a fine dispersion. A suitable quaternary amine can be reacted with the clay surfaces, which are thus changed from hydrophilic to hydrophobic (rejecting water) but oleophilic (attracting oil).

At this point, surface area is not large enough for effective thickening of lubricating oil. A polar activator or dispersant may be added. The oleophilic or organo-clay, oil, and activator are then intensively milled, breaking the clay particles to a great number of very fine platelets. When surface area is large enough, a smooth, stable grease is formed.

Organo-clay grease illustrates the need for attraction between thickener and fluid and, simultaneously, large surface area. Organo-clay greases do not develop fibrous structures, such as are formed with soap greases.

Urea Derivatives

Another important type of nonsoap grease is formed from urea derivatives. Urea, $H_2N\text{-}CO\text{-}NH_2$, is water-soluble and oil-insoluble. Through reaction, hydrocarbon chains and/or rings can be hooked onto the urea molecule, increasing oil solubility, but still not thickening effectively. Such a substituted urea can be formulated as follows:

$$RHN\text{-}CO\text{-}NHR$$

Note: The R in such a formula represents a hydrocarbon substituent.

Further work showed that if two or more molecules of these substituted ureas are hooked together, the resulting chemical may be an effective thickener for oil. Diureas and tetraureas, both identified as polyureas, have been used commercially. Note that they are not polymers. In a true polymer, whole basic molecules are hooked together. In the polyureas, only the urea portion of the molecule is repeated. This is done by reacting precise proportions of amines, diamines, and diisocyanates, according to the product sought.

The nature of the reactions which are carried out in oil leads to chainlike molecules. The chemicals chosen have some oil solubility. And reaction conditions, including intensive milling, form numerous small particles. The result is a different sort of chemical whose thickening mechanisms resemble those of soaps.

Soap Complexes

Over the years, grease technologists have sought ways to improve heat resistance of soap greases. Their efforts led to the development of soap complexes. Each soap requires a different complexing agent, but all appear to have one common characteristic: The complexing agent is a salt. Thus a soap thickener is converted to a soap-salt complex thickener.

If we measure heat resistance by dropping point, to be discussed later, the dropping point of a complex grease is at least 100°F (56°C) higher than the dropping point of the corresponding soap grease. As a significant result, complex greases are usable to higher temperatures than are effective with the corresponding soap-thickened products.

In manufacture, if the salt, the soap, and the oil are mixed together, we obtain nothing useful. Instead, the fatty acid, the nonfatty acid, alkali (or alcoholate), and oil are mixed together in a controlled cooking and mixing process. The soap and salt are formed and cocrystallized, developing the complex thickener.

The mechanism of forming such a complex is not always known. In some cases, such as aluminum complex, the metallic element appears to become attached to both the fatty acid and the nonfatty acid. Each molecule of thickener would then consist of:

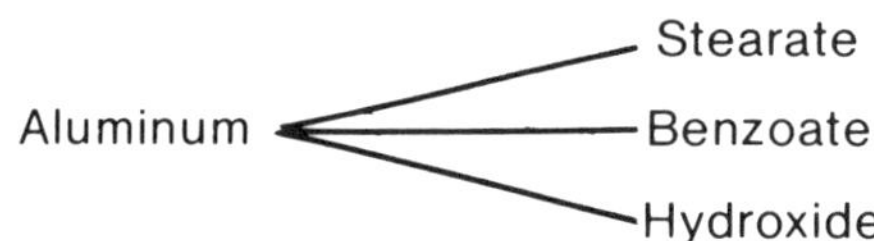

In other cases, crosslinking of soap crystals seems to take place. This seems the most likely mechanism for lithium complex greases, where the crosslinking agent is the lithium salt of a dibasic acid.

In the case of calcium complex greases, there could be several crosslinking mechanisms. Which one is effective may depend on the nature and proportions of ingredients used and processing conditions. The complexing salt for calcium complex greases is essentially calcium acetate. Calcium hydroxide and calcium carbonate may also contribute to this complex formation. The calcium complex thickener is often depicted as:

This simple formula does not seem to explain the complex chemical reactions which take place in forming calcium complex grease.

Thickener Structures--Electron Micrographs

We have mentioned fiber structures of greases. Some years ago, use of the electron microscope enabled us to see the particles, often fibers, of which the thickener structures are formed. These structures are responsible for variations in gross properties of the different products.

Figure 1.3 - Sodium Tallowate Grease

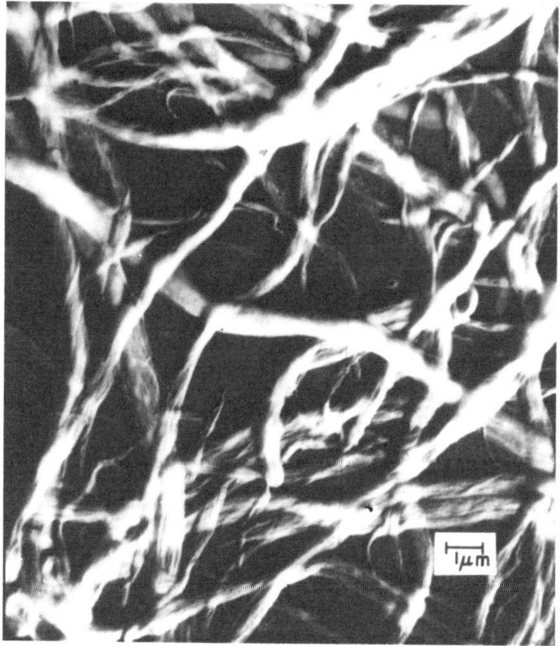

Photo courtesy of Texaco Inc.

Sodium soap greases, for example, are generally fibrous and stringy. The electron micrograph for such a product (Figure 1.3) shows long, thick, tangled soap fibers. Water-stabilized calcium soap greases, on the other hand, form relatively small fibers with little tangling (Figure 1.4). Anhydrous calcium 12-hydroxystearate greases have somewhat longer and more tangled fibers, still small enough to form smooth-textured greases (Figure 1.5).

LIthium 12-hydroxystearate greases contain longer, visibly twisted, and well-tangled fibers, still much smaller than those of sodium soap greases (Figure 1.6). The other electron micrographs shown illustrate other types of thickener particles--platelets, clumps, small fibers and small particles. Small fibers or nonfibrous structures are found in smooth greases (Figures 1.7 through 1.11).

Figure 1.4 - Water Stabilized Calcium Grease

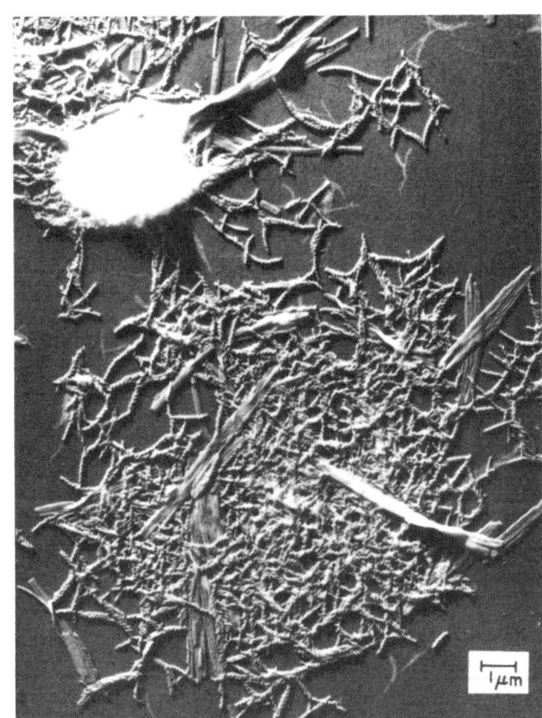

Photo courtesy of Texaco Inc.

Figure 1.5 - Calcium 12-Hydroxystearate Grease

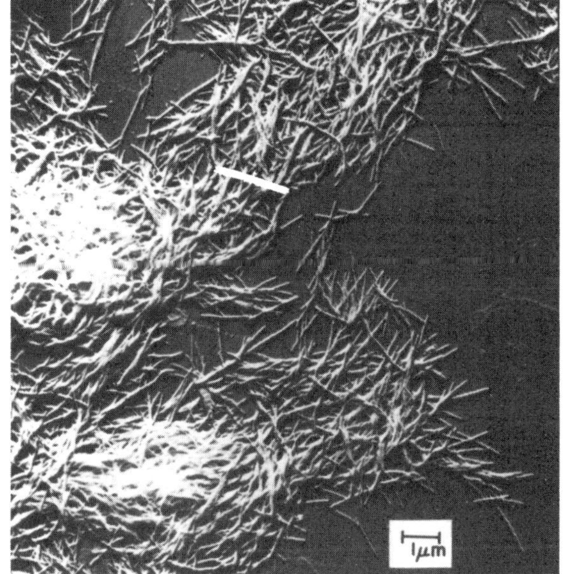

Photo courtesy of Texaco Inc.

Figure 1.6 - Lithium 12-Hydroxystearate Grease

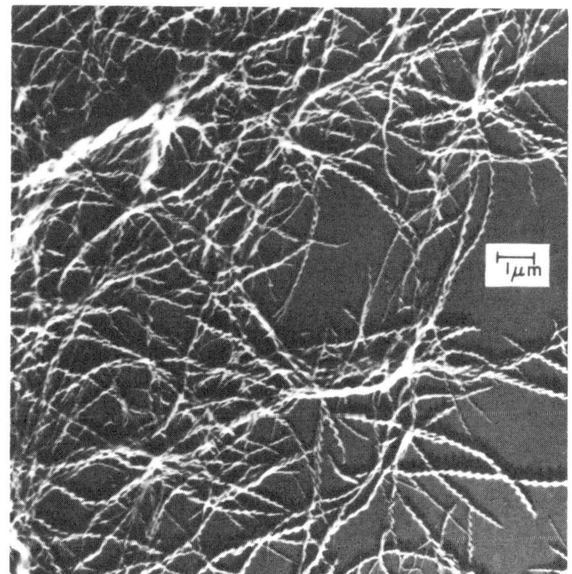

Photo courtesy of Texaco Inc.

Figure 1.7 - Calcium Complex Grease

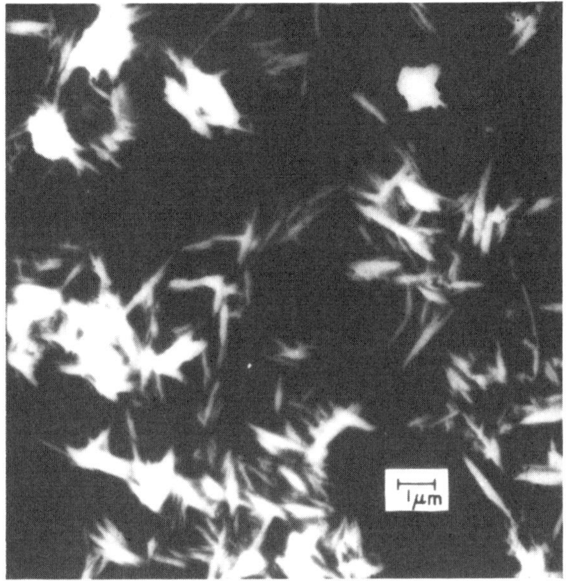

Photo courtesy of Texaco Inc.

Figure 1.8 - Aluminum Complex Grease

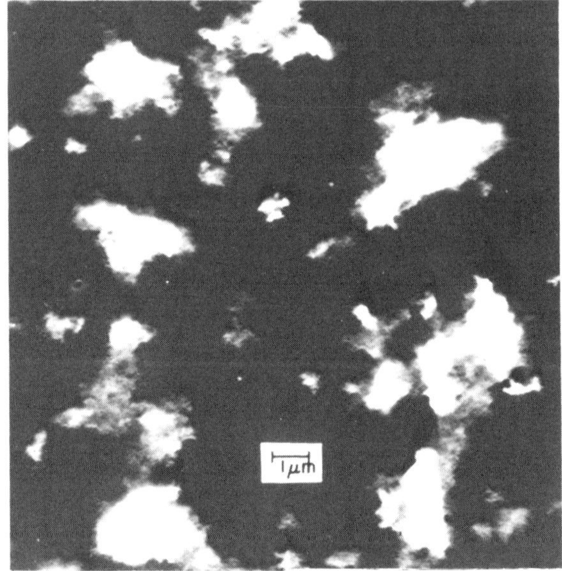

Photo courtesy of Texaco Inc.

Figure 1.9 - Lithium Complex Grease

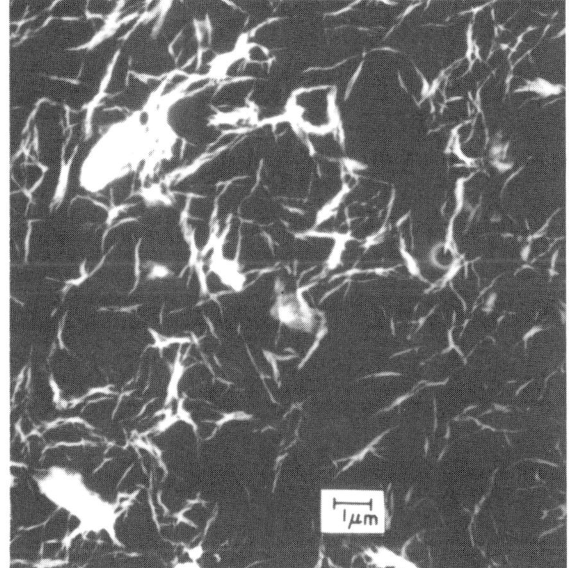

Photo courtesy of Texaco Inc.

Figure 1.10 - Organo-Clay Grease

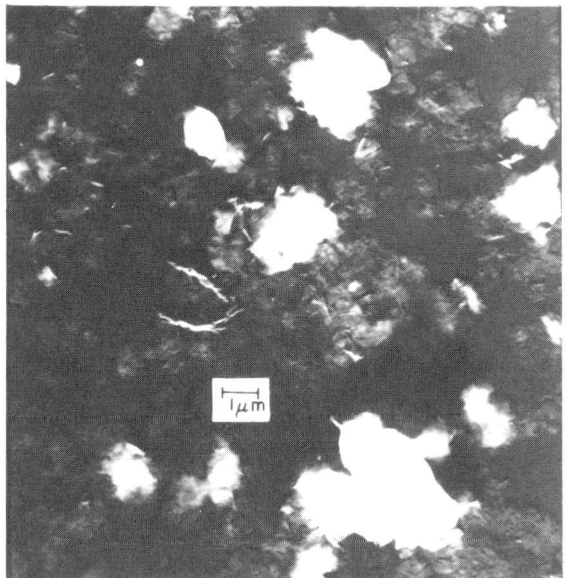

Photo courtesy of Texaco Inc.

Figure 1.11 - Polyurea Grease

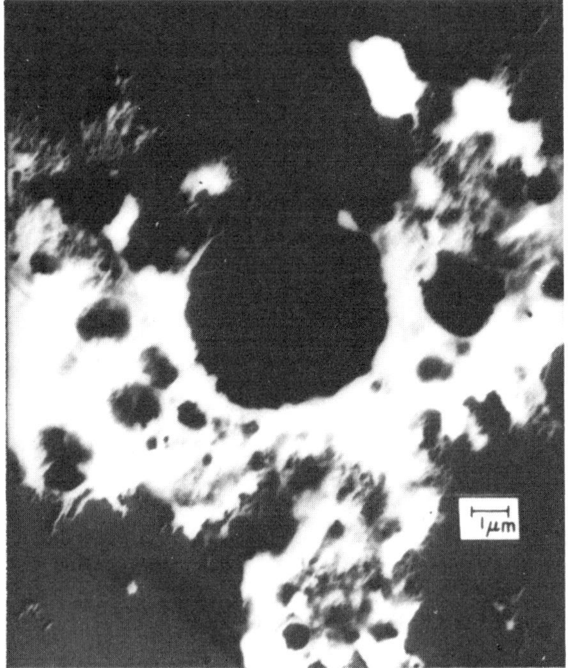

Photo courtesy of Texaco Inc.

CHAPTER 2

MANUFACTURING MODERN GREASES

In the manufacture of lubricating greases, equipment, ingredients, and processing techniques vary widely. We will describe only typical formulations and processing. The relative importance of the various types of greases is disclosed in the survey carried out periodically by the NLGI and updated and published in the *NLGI SPOKESMAN.*

EQUIPMENT USED

Grease Kettles

Most greases are processed in batches within large, essentially cylindrical open vessels. Since thickener is to be dispersed in a lubricating liquid, we must have some form of mixer. Again, since some greases are thick, or viscous, stirrers are usually powerful and heavy. Such a stirred vessel (Figure 2.1) is usually known as a grease kettle; a common capacity may be 1500 to 2000 gallons (5.7 to 7.6m^3). For lower-volume products, smaller kettles are preferred. For laboratory work, "kettles" of capacity as low as 5 pounds (2.3 kg) have been described.

Figure 2.1 - Grease Kettle

Photo courtesy of Struthers Wells Corp.

In making most greases, kettle contents must be heated, then cooled. Heating may be done on an open fire, in which case cooling is usually accomplished with cold oil, which is part of the formula. Jacketed kettles are also utilized. These are double-walled vessels, with space for a heat transfer medium in the space between the walls. A heating medium first heats kettle contents; when the target temperature is reached, the heating medium is replaced by a cooling medium, and kettle contents are cooled.

Mixing in grease kettles is essentially horizontal. To improve and speed mixing, contents are commonly pumped out the bottom and returned to the top of the kettle. The same pump can be used to empty the kettle and sometimes to transfer its contents to storage tanks. Such pumps are usually powerful and fairly slow. Multi-speed drives are common (Figure 2.2).

Figure 2.2 - Grease Pump

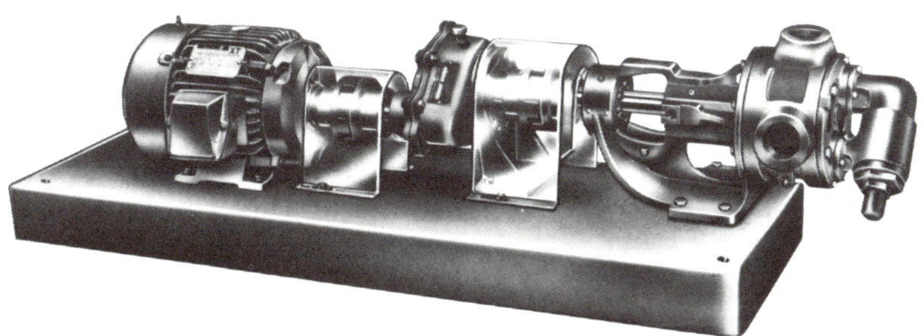

Photo courtesy of Viking Pump Div., Houdaille Industries, Inc.

As in all dispersion technology, grease thickeners must be dispersed uniformly and are most effective when particles are small. In manufacturing greases, the use of mills or homogenizers helps develop small thickener particles and improves uniformity of dispersion. However, such devices are not used on all greases.

Deaerators are also used on some products. They generally improve clarity or brightness of product, increasing sales appeal; they do not improve quality. The mixers in grease kettles sometimes beat a great deal of air into the product; it is then sometimes not possible to put the required weight of product into its container. Deaeration removes extraneous air, permitting the required weight of grease to be packaged into the container.

Filtration is used to remove undesirable foreign matter from the product. Most greases are filtered, although a few--in particular, very firm, blocklike greases--cannot be passed through fine filters. NLGI recommends that centralized grease dispensing systems contain line strainers equipped with the equivalent of 40-mesh screens (420-μm openings). For greases which may be used in such systems, filtration through 40-mesh or finer screen is obviously desirable.

Variations in Kettle Design

The thickeners in most soap-thickened and complex greases must develop fibers. Such fibers crystallize out of solutions of soap in oil or out of the molten state. The fibers may be straight or twisted; what is important is that they become entangled into a three-dimensional matrix which holds oil.

Since soaps have different melting points or solubilities in oil, processing temperatures vary. Thus we must provide for controlled heating and cooling for grease kettles and their contents.

The simplest and least costly heating system is an open fire. Many grease kettles are heated with gas. Several problems may arise. First, if the flame impinges on the kettle body, grease ingredients may become scorched, forming dark particles and sometimes an unpleasant odor. If the kettle wall temperature is above the flash point of kettle contents, a flash fire can take place. Another problem arises in cooling. The grease cannot be finished and packaged until it is cool. A fireheated kettle is not easily cooled, except by its ingredients.

Several approaches arise from these problems. Heating systems can be designed so that the burner flame does not impinge directly on the kettle surface. Scrapers are sometimes attached to what amounts to the ends of the grease paddles so that the grease will keep being removed from the heated surface. In heating, scrapers are only moderately effective but do prevent or minimize scorching.

Because of the need for cooling, only part of the oil is used during the heating process. Thus a grease "base" is formed from all the soap-making ingredients, but only part of the oil. After the reaction phase is complete, the base is dehydrated, then heated further to a temperature which varies with the formulation. Heat is turned off, and cooling is accomplished with the balance of the oil. This technique is not satisfactory with high-soap-content greases, since a great deal of oil must be incorporated into the base, and little or none is left for cooling.

Instead of a fire-heated kettle, many grease manufacturers have installed jacketed kettles. These are more expensive than the single-walled, fire-heated kettles. A baffle in the jacket directs fluid through the jacket; this baffle prevents the fluid from taking the shortest path from inlet to outlet, which would make heat transfer very poor.

Jacketed kettles eliminate the possibility of scorching and permit effective cooling. Where temperature requirements are mild, say up to about 325°F (163°C), steam is commonly used for heating. Then water may be used in the jacket for cooling. Such a system is modest in cost, but limited in temperature capability.

Steam at 100 psi (690 kPa) brings kettle contents to a temperature of about 325°F (163°C). Except where higher-pressure steam is already available, 100- to 150-psi (690- to 1030-kPa) steam is in common use in grease plants.

For most modern greases, significantly higher processing temperatures are required. Then, if a jacketed vessel is to be used, hot oil or other high-temperature media can be employed. With suitable controls, oil can be used to attain high temperatures, such as 400° to 450°F (204° to 232°C), or a lower temperature typical of steam heating, such as 300°F (149°C). For cooling, a tank of cold oil and a heat exchanger permit bringing the temperature down.

During grease processing, especially during the cooling cycle, grease becomes firmest at the kettle walls, acting as an insulator and slowing down the cooling process. In this phase, scrapers are quite effective, and even necessary.

Since grease bases are heavy and since the grease batch becomes heavier during the cooling process, simple kettle paddles will tend to move the grease around and around without significant mixing. Stationary "breaker" arms improve mixing. Even better are counter-rotating paddles, some of which rotate clockwise, the rest counterclockwise.

Alternatives to the Open Grease Kettle

We have indicated that grease kettles are usually large, cylindrical, open vessels. For the saponification reaction, however, in which fat is converted to soap, there are advantages in using a closed vessel. Here the reaction is controlled more positively, leading to more consistent product quality and shorter production cycles than are usually obtained in open kettles.

Water is required in this reaction. While heating fat, fatty acid, oil, alkali, and water in an open vessel, water is evaporated. If evaporation is rapid, lack of water may prevent the reaction from going to completion. A closed kettle can hold water in the batch. As contents are heated, water boils, raising internal pressure (this resembles the familiar pressure cooker). The same kind of mixer is used as is found in open grease kettles.

At the end of the desired time and when temperature and pressure are on target, contents of the pressure vessel are transferred to an open grease kettle, where dehydration, addition of oil, cooling, etc., are carried out. In such a scheme, the pressure vessel would likely be smaller than the open grease kettle, since only a soap base is prepared there.

Figure 2.3 - Contactor

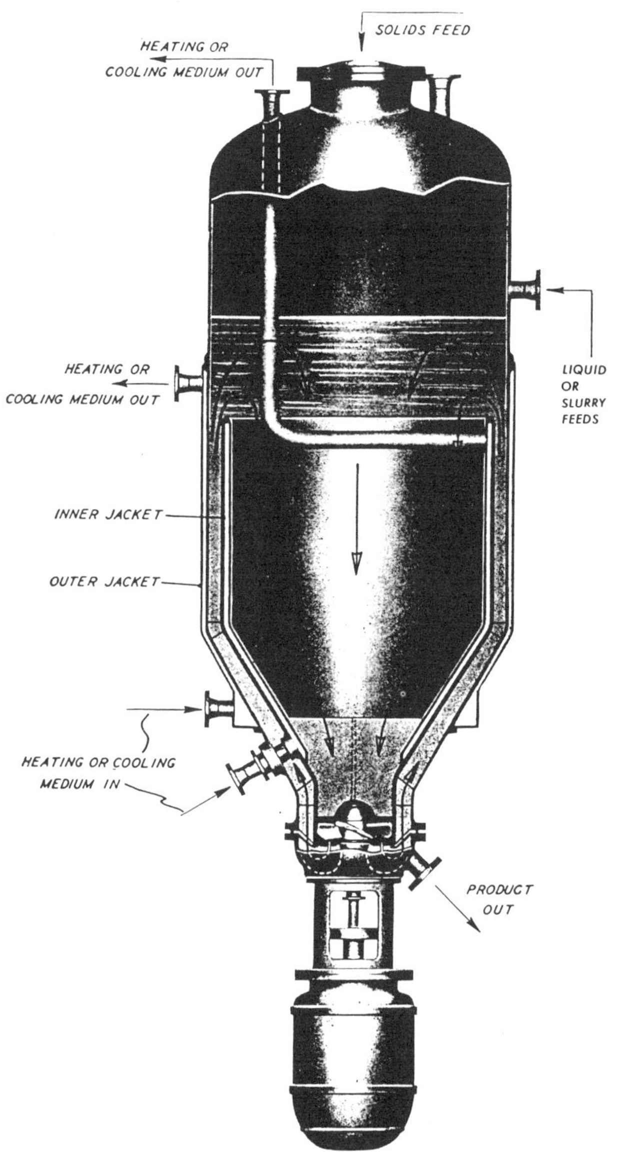

Photo courtesy of Stratco, Inc.

One common variation on the pressure cooker is a contactor. One such device is a jacketed pressure vessel of generally conical shape, having a large heating surface. In the contactor (Figure 2.3), contents are driven by a high-speed impeller rather than the low-speed heavy paddles of the grease kettle. This is generally an effective, high-speed cooker.

If the contents become very firm, the impeller may channel through the grease base, resulting in loss of circulation. This can be observed by watching motor amperage, since channeling reduces the load on the impeller motor, lowering the current drawn by that motor. Circulation can be recovered by dilution with oil or by raising internal pressure with steam or an inert gas.

Continuous Grease Manufacture

For making soap-thickened greases, a continuous process unit has been developed (Figure 2.4). Continuous processes generally are most effective when used for long runs of one product; frequent changes cause loss of efficiency and sometimes quality.

Manufacturers who make a large volume of one product or a few products have found this unit to show savings in labor and energy. Lower material cost and more consistent manufacture are also attainable. For use with other than soap greases, the efficacy of this device would have to be developed through experimentation.

Figure 2.4 - Continuous Process Unit

Photo courtesy of Texaco Development Corporation

Grease Plant Flow Diagram

A drawing showing a pressure vessel, two open kettles, and associated tanks, pumps, lines, etc. appears in Figure 2.5. This gives the overall layout of a grease plant, from raw materials to packages.

Figure 2.5 - Process Flow Diagram

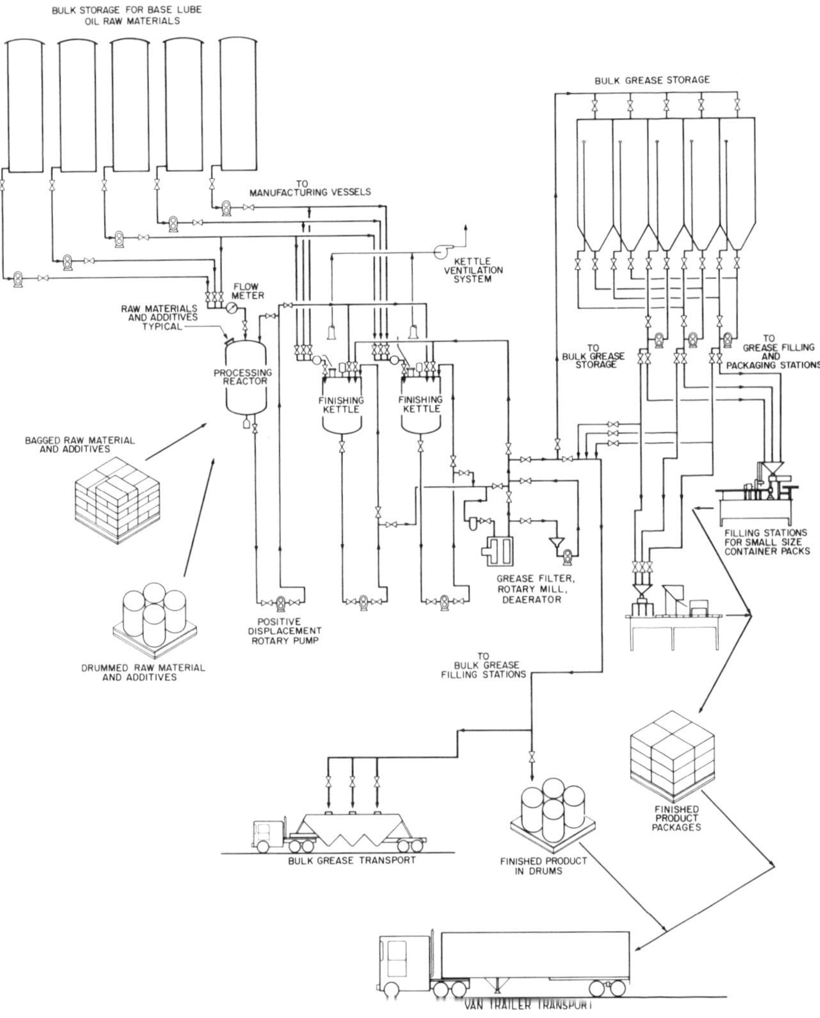

Photo courtesy of Southwest Div., Witco Corp.

SOAP-THICKENED GREASES

Grease manufacture varies with the thickener used. For ease of comprehension, it is best to separate thickeners into groups. First, we will consider soap-thickened greases. We will follow with soap complexes, organo-clay, and polyurea. Other specialized thickeners are used in small volumes and will not be discussed at length.

Aluminum Soap Grease

The simplest greases to make are aluminum soap-thickened products. Most manufacturers purchase a preformed aluminum distearate soap and place a suitable amount in a grease kettle. A typical proportion might be 8 percent of the batch, but this will vary with processing conditions, the oil used, and the consistency of product desired.

Oil is added in suitable proportions, perhaps one-third to one-half the total needed. Most aluminum soap greases are made easiest with naphthenic oils. The mixer is turned on and the contents mixed without heating until uniformly slurried. The kettle is heated, raising temperature to 280° to 300°F (138° to 149°C). By this time, the soap is dissolved in the oil. The balance of the oil is added while still heating, forming a uniform, clear, thick gel.

The hot, molten gel is usually drawn out of the kettle into pans. Here the grease is allowed to cool, without mixing, to near room temperature, forming a firm cake. If mixed during cooling, the product will become soft. Fast cooling results in firmer product; slow cooling gives softer product. After cooling, the firm cakes of grease are shoveled or scraped from the pans and pumped through a screen or working device, which makes the grease smooth and uniform.

The same grease can be made in a semicontinuous process. The soap and all the oil are slurried; the slurry is pumped through a mixing heat exchanger, then a cooling heat exchanger. Soft product flows into a storage tank, where it stands for a while, often overnight, being transformed to a firmer gel of satisfactory characteristics. This can be smoothed, as before, by pumping.

Hydrated Calcium Soap Grease

Grease formulators generally prefer not to use preformed soaps. Instead, soap-making ingredients are reacted in the grease kettle, forming the soap. Kettle-manufactured soap is often less expensive than preformed soap. A further and more important factor is the flexibility in formulation possible when the batch is started from fat, oil, and alkali.

In making calcium soap grease, the basic fatty ingredient is tallow. Many variations have been used commercially. At one time, horse fat was the fat of choice; hog fat (white grease or yellow grease) also has been widely used. Today, beef or mutton tallow is most commonly employed. Fatty acids derived from these or similar sources have also been utilized but, when used alone, make unstable greases. However, blends of fat and fatty acid are common.

The alkali needed for making calcium soap grease is calcium hydroxide, which has the common name "hydrated lime." Calcium soap greases are thus often known as "lime soap greases." The amount of lime used must be the amount calculated and a small excess, to permit the lime and fat to react completely, converting them to calcium soap. Since a reaction must take place, a small amount of water or other material to start the reaction is added.

In manufacture, then, appropriate amounts of fat, fatty acid (if used), and oil are added to a kettle. The mixture is warmed and mixed until contents are uniform. Hydrated lime and water are added, and temperature is raised to the target temperature. Here preferences vary--in part because of the equipment available, in part because of the oil used, and in part to achieve specific characteristics in the product. This heating is often continued until contents are around 300°F (149°C). A pressure kettle or contactor, if available, saves processing time.

The kettle contents, now a soap base, are checked for completeness of saponification. This is done by checking for alkalinity. Assuming that correct amounts of ingredients have been added, a normal amount of alkalinity can be established. Any excess alkali, then, means that the reaction is not yet complete. Heating and mixing are continued until alkalinity is normal; then heat is turned off, and cooling begins.

If only a small part of the oil was used in the cooking operation, a great deal of oil is available for cooling. This oil is added slowly at first, since the thick soap base may not readily accept oil. As the base becomes more fluid, oil is absorbed more and more rapidly. Oil addition is continued down to about 230°F (110°C). At this point, an operation unique to calcium soap greases is undertaken. Precise details vary with the equipment, the heating medium, and the cooling rate.

Should the cooling operation continue, kettle contents would separate into what looks like a coarse porridge. Free oil would separate from the soap concentrate, which breaks into hard curds. The soap seems to lack affinity for oil--even naphthenic oil, which has the best solvency. However, if we add water before cooling too far and if it is combined properly, the hydrated soap can be dispersed uniformly in oil.

When fat is used in the grease manufacture (rather than fatty acid), glycerin--a sticky, water-soluble chemical--is formed as a by-product. The glycerin may aid in hydrating the soap. However, the grease maker's "art" is particularly called for in the hydration step. Water must be added at the right temperature and with the correct timing. If so, the water will not foam out or just lie at the kettle bottom but will mix throughout the batch. Once mixing and hydration are complete, oil addition rate is speeded up to accelerate cooling so that the water will not evaporate out.

If all these steps have been carried out properly, the batch should be almost ready for drawing. If too warm, jacket cooling can supplement the cooling with formula oil. The batch is checked for consistency, additional oil incorporated according to the grade desired, and additives mixed in. The final grease contains water in an amount approximating ten percent of the soap content. This type of calcium grease is thus known as hydrated calcium grease.

Anhydrous Calcium Grease

Several unique formulations have led to the commercialization of water-free (anhydrous) calcium soap greases. The most successful of these involves the use of 12-hydroxystearic acid, described earlier.

If we react this 12-hydroxystearic acid with lime in the presence of oil, a calcium grease which contains no water can be prepared. Its heat resistance is somewhat better than that of hydrated calcium grease, but not as great as can be achieved with greases made from other thickeners. Thus anhydrous calcium grease is used commercially only in limited amounts.

Sodium Soap Grease

Sodium hydroxide reacts with fats and fatty acids to make sodium or soda soaps. Selected fats and sometimes fatty acids are placed in a kettle along with a small proportion of water and some suitable oil. Heating and mixing are carried out in a manner similar to what is described under "Hydrated Calcium Soap Grease." Several unique situations arise.

First, these ingredients tend to foam in heating. To avoid foaming out of the kettle, heating must be slow, and a silicone antifoam may be helpful. Then, as sodium soap is formed in the oil, it segregates into tough ropy fibers. Thus power requirements are high, and the mixer must be built ruggedly.

The grease can be cooked adequately at 325°F (163°C). However, sodium soap greases are made in a variety of structures. Contributing are: type of fat, type of fatty acid,proportions of the two, completeness of solution of soap in oil during cooking, residual moisture, presence of structure modifiers, effectiveness of mixing and of milling, cooling rate, etc. Thus the cooking temperature given is only acceptable for certain manufacturers, who may accept traces of moisture in the finished product and like the fiber structure of the grease which results.

Many manufacturers prefer to cook at higher temperatures, such as 400°F (204°C), at which point the soap will be completely dissolved in the oil, and even trace amounts of moisture will have been evaporated. On cooling, milling is still helpful to improve texture.

Cooling may be done with the remaining oil from the batch. If jacketed kettles are used, cooling medium may be pumped through the jacket to speed the cooling operation.

Lithium Soap Grease

If fat, fatty acid, some oil, and lithium hydroxide are heated as above, lithium soap grease bases are formed. These greases are best processed at temperatures above those usually attainable with steam. Therefore, open fire or high-temperature heating medium would be preferred. Hot oil is often used as a heating medium.

If the fatty material is 12-hydroxystearic acid, described above, the lithium grease formed has some unique properties which have been preferred in the marketplace. Cooking temperature is usually in the range of 375° to 410°F (191° to 210°C).

Standard fats such as tallow, and standard fatty acids such as a stearic acid, will also react with lithium hydroxide to form a lithium soap. Lithium soap grease of this type may be considered to have properties between those of sodium soap grease and lithium 12-hydroxystearate grease.

The hot, dehydrated lithium soap grease base, as with sodium soap grease, is cooled with the balance of the oil in the batch. If a jacketed kettle is used, cooling is speeded up by changing to a cooling medium in the jacket. And to improve texture and yield, an auxiliary milling device is normally employed.

SOAP COMPLEXES

Aluminum Complex Grease

This type of grease is made from a fatty acid, such as one of the stearic acids, and a nonfatty acid, --benzoic acid. Stearic acid can converted into a soap--aluminum stearate--and benzoic acid into a salt--aluminum benzoate.

Aluminum hydroxide will not react with stearic acid or benzoic acid; thus the reaction must be carried out indirectly. The chemical which was found to react is aluminum isopropoxide. A trimer of this chemical, sold under a trade name, is also capable of reacting with these acids.

As one example, using the trimer, suitable proportions of stearic acid, benzoic acid, and oil are heated until clear. Then the trimer is added and the mixture heated to 380° to 400°F (193° to 204°C). Heat is turned off and oil added gradually to maintain mixing and to cool the batch. Additives are incorporated as required, and the batch is milled for smoothness.

To a greater extent than with regular soap-thickened greases, solubilities, solvency, and molar ratios must be controlled carefully. That this can be done is indicated by the significant volume of this type of product now in service.

Calcium Complex Grease

The soap in this case is calcium stearate, the salt is calcium acetate. This grease can be made in pressure vessels or open grease kettles.

A wide range of products of this type has been developed. The acetate is formed when acetic acid reacts with lime while the grease base is being formed. Greases are produced with different proportions of acetic acid and fatty acid. In some cases, the fatty acid is modified with medium molecular weight acids, such as caprylic or capric acid.

This type of grease is quite sensitive to alkalinity. In manufacture, a slight excess of lime is acceptable, but a large excess may cause the batch to be soft.

Water is also a factor with these greases. They must be dry; if water is retained, they may change consistency in storage.

One other consideration: Calcium complex grease is most easily made with high thickener (complex) content--20 to 30 percent is not uncommon. However, with sufficient sophistication in formulation and processing, grease of significantly lower thickener content will produce calcium complex grease of the same consistency.

In an open kettle process, fat, fatty acid, and oil are mixed and warmed only until the mixture has become a clear solution. Lime is added and mixed vigorously, making a slightly thickened, creamy fluid. Additional heating is not desirable to this point. Glacial acetic acid is introduced in suitable proportions and with vigorous mixing. Evaporation and fumes of acetic acid should be minimized. As the acetic acid is converted to calcium acetate, its odor disappears. The batch is heated and mixed until dry. In order to remove all traces of water, cooking at a higher-temperature such as 400°F (204°C) is required.

The grease base is then cooled and diluted with oil, additives are incorporated, and the batch may be milled until smooth. Milling must be controlled carefully, since the batch may become quite firm during milling, requiring oiling. If thickener content falls too low, the product loses stability.

With all these complications, the need for sophistication in manufacture is apparent. Yet a large amount of this type of grease is still being used, although the volume has been dropping.

Since a critical reactant, acetic acid, is quite volatile, cooking in a pressure vessel is often preferred. This material can be added just before the vessel is closed or introduced into the closed vessel. Such an approach decreases exposure of manufacturing personnel to the unpleasant and hazardous fumes.

Barium Complex Grease

Minor amounts of grease are made from barium soap. In most current production, the thickener is not a normal barium soap, but rather a barium complex using acetic acid, again, as complexing reactant. Manufacturing technique can be similar to that described for calcium complex, with a significant difference in power requirement, since the barium complex base becomes very firm and stringy during manufacture.

Lithium Complex Grease

There are numerous variations in the manufacture of this product; only one will be described. As fatty acid, we can use 12-hydroxystearic acid. As complexing agent, we use a dibasic acid, often more conveniently introduced as the dimethyl ester. Thus dimethyl azelate or sebacate has been used. A common procedure is carried out in two steps: First the normal soap is formed, then it is complexed.

In an open-fired, open grease kettle, the fatty acid and oil are warmed until clear. Lithium hydroxide is added, and kettle contents are heated until the reaction forming normal soap is complete. More lithium hydroxide is now required, since it must react with the dibasic acid. The full amount of lithum hydroxide needed for both reactions is often added at the beginning, the excess from the first reaction remaining available for the second reaction.

The selected dibasic acid ester is added and heating and mixing continued to a chosen temperature, which could be 400°F (204°C). At this point, complexing should be complete and the batch dehydrated. Oil is added for cooling (jacket cooling is also used if the kettle is jacketed) and additives incorporated. The batch is milled and cooled for drawing.

The addition of ester into the hot grease requires caution. Some esters have high volatility and low flash points. When added to hot grease, a flash hazard may be created. Cooling the batch before adding ester may be desirable. Alternatively, blanketing with an inert gas in a closed vessel may be preferred. In actual manufacture over many years, no flashbacks have been experienced, but the possibility must be recognized.

NONSOAP GREASES

There are, at present, two major types of nonsoap greases. One uses an insoluble powder which is introduced into the grease without any reaction. The other is an organic chemical of the urea series, which is formed in the grease kettle through reaction. Thus each is a unique type of product.

Many lower-volume thickeners have also been made over the years. Several of them deserve brief mention.

Organo-Clay Grease

The organo-clay must be chosen from the several available, with consideration of the oil to be used. Grease production requires thorough dispersion of the organo-clay in oil. In one technique, all the organo-clay for a batch and some of the oil--usually one-half to one-third the formula amount--are loaded into a grease kettle and mixed until uniform. A special pregelling agent, often called a polar dispersant, is added--about one-fourth of the amount of organo-clay used. Mixing continues for about 30 minutes. Additives may be incorporated here along with the balance of the oil. The batch must then be milled. Intensive milling is necessary.

Several variations in the manufacturing process are available if we want to minimize cost and maximize uniformity and quality. The polar dispersant or pregelling agent actually affects swelling of the clay particles. Several pregellants are common: acetone, ethanol, methanol, and propylene carbonate. Some water is essential to the swelling process. A small but variable amount of water may be present in the organo-clay and the pregellant. Experimentation should permit the selection of optimum proportions of pregellant and water with relation to the clay and oil used.

Hazards must be recognized. Acetone, methanol, and ethanol are quite volatile and flammable. They form explosive mixtures. All must be handled with care. Breathing their vapors should be avoided.

The odor of volatile pregellant is often noticeable in the finished product. Many find this unpleasant. In addition, enough of such volatile material may accumulate in a container to create a flash hazard, should the container be opened near an open flame or glowing cigarette. Thus some prefer to heat the unmilled mixture, prepared as above, to about 250°F (121°C) until free of odor (and water). It is then cooled below 180°F (82°C). Next, 0.1 percent water (based on the entire batch) is added. Any remaining additives and oil for adjustment are introduced, then the batch is subjected to highly effective milling. This addition of water replaces the water lost in the heating step.

Polyurea Grease

In order to make polyurea grease, oil is mixed with suitable amines in a grease kettle, an isocyanate or a diisocyanate is dissolved in oil in a separate vessel, and this solution is slowly added to the grease kettle with mixing.

The reaction proceeds rapidly, generating some heat which is controllable. There are no by-products, such as water or methanol, which are formed in the manufacture of some other greases. The mixture is heated to a temperature which depends on the chemical reactants. Oil is added for cooling, additives are incorporated along with additional oil to obtain the desired grade, then the product is milled.

A considerable hazard must be recognized. Some amines are fairly hazardous. Isocyanates and diisocyanates are considered extremely hazardous, even in remarkably low concentrations. Personnel therefore must be protected from handling or breathing these materials.

Specialty Thickeners

Many more thickeners have been used and are being used commercially, but in relatively small amounts. To indicate their range, a few will be mentioned here. These are simply mixed with fluid in a grease kettle, then passed through an effective mill or homogenizer.

Silica--Silica is found in nature as sand. However, a very fine fumed silica thickens many fluids to form high-melting greases. The silica greases are inherently sensitive to water.

Silicone fluids are water-resistant. When silica is used to thicken silicone fluids, water-resistant, high-melting (or nonmelting) greases are formed. With petroleum oils, oil, silica, and sometimes a surface-active chemical are mixed together, then milled.

As a modification of the above, one grade of silica has been treated with a silicone, making it water-resistant. This rather costly grade has been used to thicken petroleum oils and, again, make high-temperature, water-resistant greases.

Carbon Black--Some very fine carbon blacks have considerable thickening power. The resulting greases are high-temperature products but have surprisingly poor water resistance. These thickeners, again, are sometimes used to thicken silicone fluids, forming water-resistant products. They have also been used, along with soaps, to make mixed-base greases. Some users find the intense blackness of these products objectionable.

Pigments--Several pigments, available in very fine particle size, thicken specialty fluids, such as silicones. Some of these products have been used by the military. The best-known pigment has been copper phthalocyanine, which has an intense blue color and is used in printing inks.

INCORPORATION OF ADDITIVES

Many of the additives used in greases are sensitive to heat and will become ineffective at cooking temperatures. They may also be destroyed by the strong alkalies used--calcium, sodium, or lithium hydroxide. Normal practice, then, is to incorporate additives late in the cooling phase, when temperatures are below 185°F (85°C).

Some exceptions to this pattern are found, however. Some additives are not affected by heat or alkalies and can be introduced at any time. Some will not dissolve and mix properly unless added hot. Each of these materials must be checked in advance to be certain that, at the temperature added, the properties an additive is meant to provide will be found in the finished batch.

DO GREASE PLANTS POLLUTE?

Grease manufacturing plants can pollute both water and air. With due engineering consideration, plants can be designed to avoid such pollution, although at significant cost. We should look at some of the sources of pollution.

When tankers are unloaded, oil may drip during connection, during disconnection, or when the connection is not tight. Such oil gets on the ground and is washed away by repeated rains, polluting streams.

In the operation of a grease plant, materials often drip or pour on the floor. If the material is washed off, the contaminants may get into the drains. These may lead, directly or indirectly, into a body of water.

Grease kettles must be vented. This leads to several air pollution problems. Acetic acid, rancid fats, and some sulfur compounds, for example, are quite unpleasant and can get into the vent system and be blown around the neighborhood. During cooking, water (as steam) is boiled away, carrying with it vapors or aerosol particles of fat, oil, soap, glycerin, methanol, etc. Furthermore, while handling solids, such as graphite, clay, molybdenum disulfide, etc., these finely divided solids float in the air and are exhausted through ventilation fans.

Many means of controlling such pollutants are used, and they must be engineered carefully. To handle oil in the soil or in sewage effluent, dikes can be built around the polluted area and runoff led to oil separators. Oil-free water is then the effluent which runs into sewers or streams.

To avoid air pollution, vent fans draw dusts through filters, from which solids can be dropped into a container, which is sealed for disposal. Grease kettle fumes, being wet, must be handled somewhat differently. The vapors, again, can be drawn from the kettles through ducts to a wet scrubber. Liquids are cooled, condensed, and drawn off. Solids and gases leaving the scrubber go through filters, which hold any particulates. These are washed off and disposed of according to the nature of the residue. Some engineers prefer to send the dried fumes to burners.

So to answer our question, modern, well-engineered grease plants are not polluters.

CHAPTER 3

TESTING LUBRICATING GREASES

Greases are used because of their properties. Grease properties are defined and measured by a number of tests, many of them unique. Producers and consumers who recognize a gap in testing often develop new tests. Some of them become widely used; others are found in only a few laboratories. We will describe here many of the widely used tests. Grease specifications are meaningful only when their reader is familiar with grease tests and their significance and limitations.

Most of these tests have been studied and standardized by ASTM committees. They have appeared in Volumes 23, 24, and 25 of the Annual Book of ASTM Standards. Starting in 1983, the ASTM book has been renumbered, and grease tests are found in Section 5, Volumes 05.01,05.02, and 05.03. Numbers for the individual tests have not been changed. For details on exactly how these tests are run, the ASTM volumes should be consulted.

CONSISTENCY

Of all grease properties, the first one that comes to mind is consistency, or firmness of the product. Greases are available in consistencies ranging from almost fluid, or semifluid, to firm blocks. Consistency of product must be appropriate for the application. A grease that is too hard may not feed adequately into areas to be lubricated. However, if a grease is too soft, it may leak away from the area where it is needed.

Figure 3.1 - Grease Penetrometer and Cone

Photo courtesy of Texaco Inc.

Consistency also influences pumpability, softer greases generally being pumped more easily. When equipment is to be greased through a dispensing system, the consistency used may be a compromise between that required for lubrication and that required for dispensing.

Penetration

Consistency is commonly measured by the ASTM cone penetration test (D217). Penetration is the depth, in tenths of millimeters, to which a standard cone sinks into the grease under prescribed conditions. Thus higher penetration numbers indicate softer greases, since the cone has sunk deeper into the sample.

Unworked penetration is measured when a sample of grease is brought to 77°F (25°C) and transferred to a standard cup; its surface is smoothed and the cone, in its penetrometer assembly, placed so that its tip just touches the level grease surface. The cone and its movable assembly, weighing 150 g (0.33 pound), are permitted to rest on and drop into the grease for exactly five seconds. The distance dropped is measured (Figure 3.1).

Many greases change significantly in consistency when manipulated. A **worked** penetration is thus considered more significant as to service behavior than is unworked penetration. For this test, the grease is churned 60 round-trip strokes in a standard worker (Figure 3.2), again at 77°F (25°C). Air is driven out of the sample, its surface is smoothed, and again the penetration of the cone is measured.

Figure 3.2 - Grease Worker

Photo courtesy of Mobil Oil Corporation

NLGI Consistency Numbers

Based on worked penetration, the NLGI has established consistency numbers ranging from 000 to 6. These are shown in Table 3.1. Each consistency number has a range of 30 points, and between numbers, we find unallocated spaces of 15 points. Since experience has shown that our ability to repeat penetration values is about 5 points, the spaces were established in consideration of test precision. Consistency numbers have sometimes been referred to as "grades."

It is essential to remember that consistency numbers become higher with firmer greases, but penetration values become lower. The dichotomy of systems can become confusing when we deal with maximum or minimum values for consistency or penetration.

For some field problems and in certain tests, the amount of grease available is too small for standard penetration equipment. For such cases, one-half scale and one-quarter scale equipment described in ASTM D1403 can be used.

TABLE 3.1

CLASSIFICATION OF GREASES BY NLGI CONSISTENCY NUMBERS

NLGI Number	ASTM Worked Penetration
000	445-475
00	400-430
0	355-385
1	310-340
2	265-295
3	220-250
4	175-205
5	130-160
6	85-115

The most commonly used consistency number is NLGI No. 2. Softer grades, especially 0 and 1, are often used for improved pumpability or low-temperature service. Higher consistency numbers are used for certain high-speed bearings and where leakage and sealing are particular concerns.

Figure 3.3 - Roll Stability Test

Photo courtesy of Mobil Oil Corporation

Consistency Stability

Greases may change consistency in service, primarily because of changes in the size and dispersion of thickener particles as a result of mechanical shearing. The resistance to such change is referred to as consistency stability, shear stability, or mechanical stability. Depending on the particular grease and application, greases may harden, soften, or go through a more complex sequence of hardening and softening. It is important that a grease have consistency stability adequate for the shear to be imposed on it in service and for the expected service interval between relubrications.

Consistency stability is commonly measured by prolonged worked penetration tests (D217) or by the roll stability test (D1831). It should be noted that these tests are at relatively low shear rate. While of some use in many common applications where shearing is not intense and relubrication is reasonably frequent, they are not usually predictive of shear-intensive service with very infrequent relubrication. For such service, special testing may be required.

In the roll stability test (Figure 3.3), a cylindrical roller weighing 11 pounds (5 kg) is inserted in a chamber along with 50 g of grease. The chamber revolves at 165 rpm for two hours at room temperature. This motion causes the roller to mash, knead, and otherwise work the grease in the chamber. Worked penetrations are taken before and after rolling; because of their small sample size, these penetrations are determined with the one-fourth scale or one-half scale worker (D1403).

FLOW PROPERTIES

Viscosity and Apparent Viscosity

Although some grades of lubricating grease are firm, in order to be dispensed and to lubricate, they must flow. Flow requires stress; the resistance to flow of a liquid is called its viscosity. Liquids such as additive-free lubricating oils are known as Newtonian fluids in that shear rate is proportional to the pressure or stress applied. This proportion is the coefficient of viscosity, usually called **viscosity**:

$$\text{Viscosity} = \frac{\text{Shear stress}}{\text{Shear rate}}$$

Greases are not Newtonian. When stress is applied, they do not flow until a level of stress known as **yield point** is exceeded. Then flow proceeds, increasing as stress is raised, but not in a proportional or linear relationship. In order that this nonlinear response not be confused with Newtonian viscosity, the viscosity relationship for greases under flow is called **apparent viscosity**.

The difference between oil viscosity and grease apparent viscosity can be summarized this way: Apparent viscosity of grease varies with temperature and shear rate; viscosity of oil varies with temperature alone.

Apparent Viscosity

D1092, "Apparent Viscosity of Lubricating Greases," describes a method of measuring this characteristic. In this test, the grease, at a specified temperature, is forced from a pressure cylinder through a capillary tube by a floating piston actuated by a gear-pump-driven hydraulic system. Hydraulic oil pressure is measured. Eight different capillary tubes and two pump speeds are used, for a total of 16 shear rates. Results are usually presented as a plot on log-log paper of apparent viscosity versus shear rate.

This information can be valuable in predicting the ease of pumping and dispensing grease, in estimating the pressure drop in a distribution system, or in determining the optimum pipe diameters for such a system. A group of charts available from the NLGI simplifies the conversion from the apparent viscosity-shear rate curve to pipe flow data. (One such chart is shown in Figure 3.4).

Mobility

A mobility test developed by United States Steel is somewhat similar to the apparent viscosity test, since it employs the same pressure cylinder and one of the eight capillary tubes. In this test, however, the piston is activated by pressurized nitrogen, and the grease is collected over a given time and weighed to determine unit flow measured in grams per second. These measurements can be run at various temperatures. This test is being used to predict the pumpability characteristics of greases at low temperatures.

Low-Temperature Torque

In order for a greased bearing to be lubricated, the grease must flow as the bearing turns. In forcing the bearing to move through the grease, resistance is developed and shows up as an increase in torque over what would be measured if the bearing had been lubricated with a light liquid, such as kerosene.

Figure 3.4 - Pipe Flow Chart

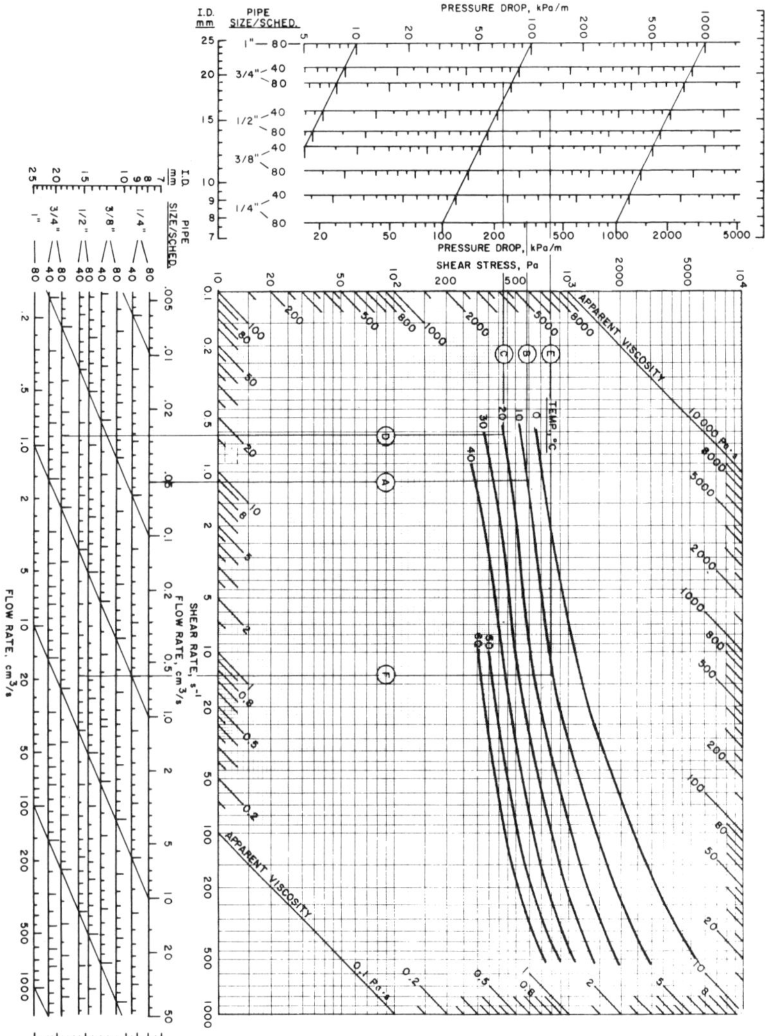

Most machines are supplied with much more power than is needed, so torque is not a significant consideration. However, some devices are low in available power, whereupon torque caused by a grease lubricant is an important consideration. Even more so is the use of such lightly powered equipment at low temperatures, as occurs, for example, with aircraft control devices. Here, low-temperature torque can be a limitation to the use of the equipment.

In D1478, low-temperature torque of a grease is measured at -65°F (-54°C) in a ball bearing. Other test temperatures may be used. After cooling statically, the motor is started, turning the bearing at 1 rpm. A cord wound around the bearing housing and attached to a dynamometer restrains rotation of the bearing with a force shown on the dynamometer. After a few seconds, a maximum reading is taken and calculated as starting torque. As rotation continues, torque falls. After 60 minutes of operation or some other agreed-on time, another reading is taken and calculated as running torque.

The method has poor precision, possibly in a large measure because of the difficulty in precisely holding and measuring such a low temperature. However, use of the results has proven helpful in design of devices for aerospace applications.

A similar method, using a larger roller bearing and run at -40°F (-40°C), is being studied by ASTM. It too is troubled by poor precision, which has delayed acceptance of the procedure.

HEAT RESISTANCE

Heat affects greases in a number of ways. When heated, grease generally becomes softer and flows more readily. At elevated temperatures, oxidation is more rapid, oil separation ("bleed") increases, and oil evaporates significantly. All these phenomena can be tested to indicate whether a product is likely to be satisfactory at temperatures encountered in service.

Dropping Point

Grease does not have a sharp melting point. Instead, on heating, it becomes softer until at some point it no longer functions as a thickened lubricant. While being heated, grease may soften gradually or sharply. A common test called "dropping point" tries to measure the end point of this process. Grease is packed into a standardized thimble or cup, which has a standard hole at the bottom.

Figure 3.5 - Dropping Point Test

Photo courtesy of Southwest Div., Witco Corp.

Most of the grease is removed with a straight rod, a thermometer inserted, and the cup placed in a standard assembly in a test tube.

There are actually two dropping point procedures. In the older one, the test tube assembly is heated at a prescribed rate in a beaker of oil. This test is ASTM D566 and is useful as long as the oil in the beaker does not flash, causing fire and burn hazards. The slow heating rate prescribed makes this a time-consuming test when used with high-temperature greases.

The hazard and slowness of the old procedure prompted modification of the method. An aluminum block replaced the beaker of oil, and test time was reduced by preheating the block to fixed temperatures. This improved method is identified as D2265 (Figure 3.5). In order to make results with the new method match results with the old method, a correction factor was built into the calculations of D2265.

In carrying out the test, the aluminum block is preheated to a temperature which depends upon the expected dropping point of the grease sample. The test tube-cup assembly is dropped into a hole in the block, and the cup is watched. At some temperature, a drop comes out of the cup and falls to the bottom of the tube. The sample temperature is read immediately, as is the block temperature. The dropping point is the sum of sample temperature and one-third the difference between that and block temperature.

Note that we do not identify the drop. Some grease structures become unable to hold all their oil and thus separate or bleed a drop of oil. Regardless of its nature, that drop defines the dropping point.

One caution should be observed. Dropping point is far above the highest temperature at which we may use a given grease. Performance capabilities should be measured by a performance-type test or actual experience.

Thickener Melting Point

Some grease specialists like to determine melting point of the grease thickener as extracted from the grease. They reason that above its melting point, the thickener cannot carry out its function. Most grease people feel, however, that the whole grease system may become too thin to be satisfactory well below the thickener melting point. They thus prefer to test the whole grease, using dropping point, trident probe, or some other test for heat resistance.

Trident Probe

Flow at high temperature can be determined with a trident probe in ASTM D3232 (Figure 3.6). The probe is mounted in the clutch of a Brookfield viscometer. The grease sample is placed in an annular chamber cut into an aluminum block.

Figure 3.6 - Trident Probe Test

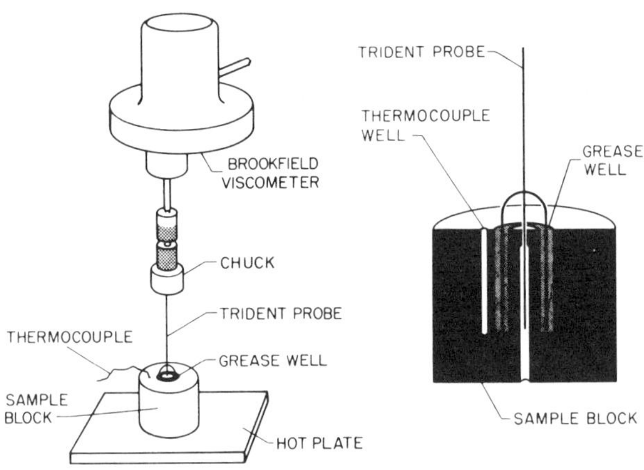

Photo courtesy of Mobil Oil Corporation

The block is placed on a hot plate and heated at a uniform rate while the trident probe rotates in the grease. The Brookfield measures torque at steadily rising temperature. When calibrated, these readings are converted to apparent viscosity. We thus obtain curves of flow at different temperatures. Observations must also be made during heating, since foaming, thickening, and similar behavior are part of the uniqueness of different types of greases, which help in identifying those greases as well as in establishing their limitations for service.

The usefulness of the trident probe is apparent when we look at the two types of drops which fall at the dropping point. If the whole grease becomes soft and drops out, the trident probe curve will show low apparent viscosity at that temperature. If a drop of oil alone falls at the dropping point, the rest of the grease should still be firm. The trident probe curve at that temperature will show high apparent viscosity. The trident probe test thus gives information not obtainable in the dropping point test.

Oil Separation

We have earlier noted that thickener is not soluble in the oil it thickens but must have some attraction for that oil. With a large proportion of thickener, that attraction is strong. As more oil is mixed in or thickener content decreases, the forces between thickener and oil decrease. Thus some of the oil in a grease may be loosely held and easily separated.

Even when stored at room temperature, some of that oil will separate. The visible separated oil is known as bleed. An ASTM test, D1742 (Figure 3.7), determines the amount of oil likely to bleed out of a grease stored around room temperature in a 35-pound (15.9-kg) pail. Storage in other containers should give similar results. That loosely held oil is part of the lubricating mechanism in grease lubrication, since a bearing lubricated with a dry (nonbleeding) grease is apt to be noisy in service.

Figure 3.7 - Oil Separation Test

Photo courtesy of Southwest Div., Witco Corp.

As grease is heated, bleed increases. No ASTM test has been standardized to measure this behavior. A test does appear in Federal Test Method Document 791, Method 321, describing a cone bleed test. Variations of this test are widely used. Grease in a 60-mesh wire screen cone is suspended in a beaker and kept in an oven; 50 hours at 212°F (100°C) is one common set of conditions. If bleed is rapid in this test, life of grease in a bearing operating at test temperature will be short. But low bleed does not, in itself, assure long bearing life.

Evaporation

Evaporation of a grease or the oil component of a grease is not a serious problem in most applications. At common bearing operating temperatures, most commercial oils evaporate so slowly that this property has not caused concern. However, some consumers, wanting assurance that greases they purchase will not be made from volatile oils, desire to have an evaporation test run.

Greases designed for low-temperature service or for both high- and low-temperature service must be made with low-viscosity oils, often synthetic fluids. Such greases are used in military and other applications. Here evaporation becomes a limitation and must be determined.

In one procedure, the cone bleed test described above has been modified to determine total weight loss during the test. That weight loss would be due to evaporation. This technique is neither precise nor highly discriminating, since air flow is limited.

Figure 3.8 - Evaporation Test

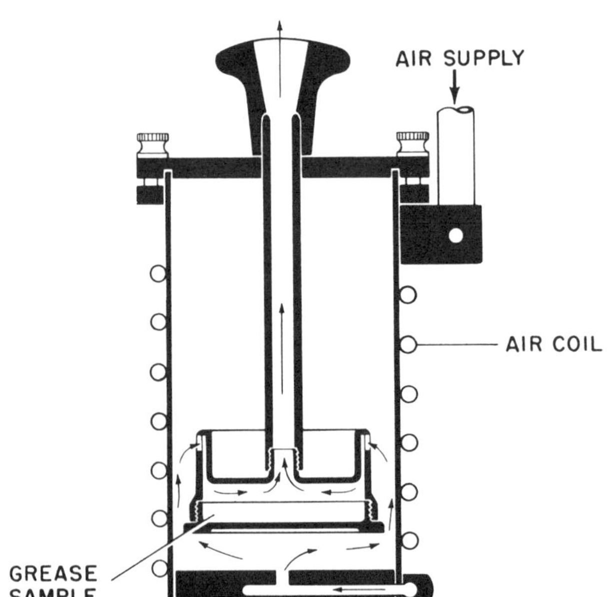

Photo courtesy of Mobil Oil Corporation

As a more accurate measure of evaporation, ASTM has standardized two methods. The older one, D972 (Figure 3.8), holds grease in a chamber which is heated to a controlled temperature between 210°F (99°C) and 300°F (149°C). Preheated air, at 2 liters per minute, is directed so as to pass over the surface of the grease for 22 hours. Weight loss is then the measure of evaporation.

As service conditions increased in severity, a modification of D972 was developed. This method, D2595, uses the same grease chamber but operates at a greater temperature range of 200°F (93°C) to 600°F (316°C). Instead of holding the grease at test temperature, but leaving air temperature uncontrolled, the new method holds exit air at the target temperature.

Those who are equipped to run D2595 have no need for D972 but should be cautioned that evaporation in the two units may not be identical. Air passing over the grease surface in the D2595 equipment is apt to be warmer than air passing over the grease surface in D972 equipment.

STORAGE STABILITY

Lubricating greases generally change in prolonged storage. If the change results in product falling out of specification range or causes performance to be below standard, that change is noteworthy and can be disturbing. One such change--oil "bleed"--has already been noted.

Another storage change is due to an inherent property known as thixotropy--a complex of work softening and age hardening. When grease is made, it is mixed and sometimes milled. This manipulation may cause softening. Then, on standing in storage, that grease may harden.

The degree of thixotropy varies widely among products. Unfortunately, grease which has hardened in storage is sometimes rejected as being too firm, whereas normal manipulation, such as would occur in applying the product, could soften it back into range.

A rarer storage change is due to low humidity. Some greases require a bit of water. When stored in desert climates, that water may be evaporated, causing the product to soften. Reintroducing a trace of water would make the product firm again.

A more severe cause of storage deterioration is due to oxygen, an active chemical which causes many products, including lubricating greases, to deteriorate in storage and in use. This will be discussed in more detail.

OXIDATION STABILITY

Storage Stability--Bomb Oxidation Test

We need to know whether oxidation will cause a grease to deteriorate severely in storage or in use. Unfortunately, no relationship between storage oxidation and service oxidation has been found. We do have an ASTM test for the former--D942 (Figure 3.9).

In this test, five glass dishes, each holding 4 g of grease, are placed in a rack inside a sealed "bomb." The grease is exposed to oxygen at 210°F (99°C) starting at 110 psi (758kPa). As oxidation of the grease progresses, oxygen is absorbed, and pressure in the bomb drops. The limit for any one grease is usually expressed as pressure drop in a given time--for example, 5 psi (34kPa) in 100 hours.

This test is widely used but has been severely criticized, since results do not appear valid for all products. It seems likely that D942 can continue as a viable test if specific limitations are written into the test method stating that (1) the test is not applicable to all products and (2) its use should be confined to products for which both producer and consumer agree that results are valid.

Oxidation Stability in Service--Bearing Tests

Even if D942 continues in use, we must remember that it has no relation to oxidation in service. No dynamic oxidation test has been approved by an ASTM committee, although such a test has been described in the literature. Most grease technologists believe that the only practical measure of service oxidation would be obtained in an actual operating bearing. There we find differences in bearing size, configuration, metallurgy, operating speed, and operating temperatures.

ASTM describes a number of bearing tests for greases. For ball bearings, we have the older and milder D1741, which uses a 306 bearing operating at 125°C (257°F) and 3500 rpm, with a thrust load of 178 N (40 lbf). The tester is operated cyclically--20 hours on, 4 hours off, until failure.

A more recent and quite severe test is D3336 (Figure 3.10). This appeared for a long time in a compilation of federal test methods but has now been standardized within ASTM. A smaller bearing, size 204, is used, operating at any desired temperature up to 700°F (371°C) and a speed of 10 000 rpm. Thrust loads may vary from 22 to 67 N (5 to 15 lbf) and added radial load from none to 22 N (5 lbf). Again, the tester is run cyclically until failure is recognized. For the higher temperature ranges, bearings of special metallurgy are required.

Figure 3.9 - Bomb Oxidation Test

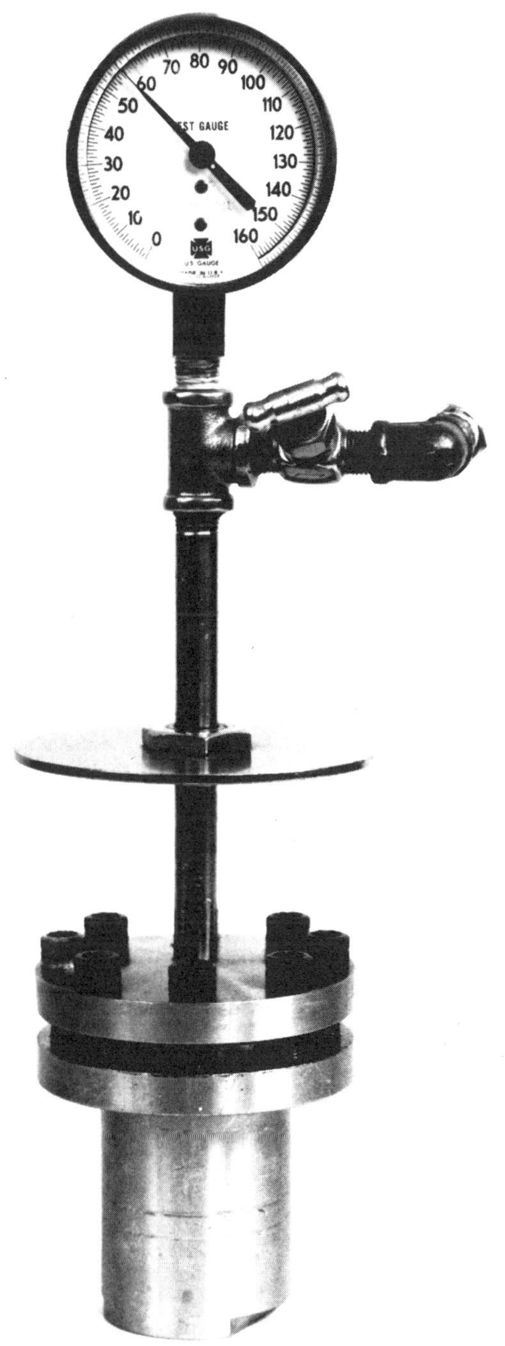

Photo courtesy of Southwest Div., Witco Corp.

Figure 3.10 - High Speed Bearing Test

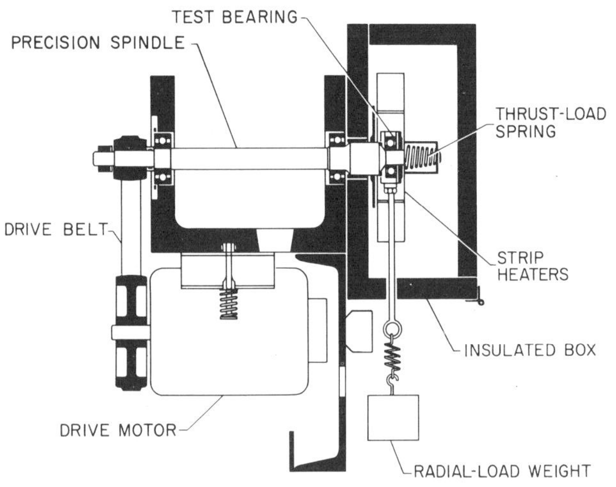

Photo courtesy of Mobil Oil Corporation

Another bearing test is D3337, which deals with greases suitable for very small bearings. This is a quite specialized use and will not be discussed here.

For automotive service or, more broadly. for use in tapered roller bearings, two service-related tests are available. The older one, D1263, runs for six hours at 660 rpm and 220°F (104°C) without load. It measures leakage but is too mild to be employed as an indication of oxidation stability. The newer method, D3527 (Figure 3.11), is much more severe. Failure in this test comes early if the grease used is not stable to oxidation.

Figure 3.11 - Wheel Bearing Test

Photo courtesy of Koehler Instrument

In D3527, two tapered roller bearings of a size to fit some passenger cars are installed on a spindle in a hub, such as are used in a compact car. The bearings and hub are lubricated; the bearings, thrust-loaded to 111 N (25 lbf), are rotated at 1000 rpm and 150°C (302°F). Operation is cyclical until failure is recognized.

Additional bearing tests are described by standardization societies of the UK (IP) and Germany (DIN) as well as bearing manufacturers, such as SKF and Ransom and Marles. It is important to recognize that grease behavior in all these tests involves--in addition to oxidation stability--mechanical stability, initial grease consistency and consistency changes, grease texture, and oil separation--both in operation, where centrifugal forces are severe, and while the tester is stationary during the off cycles. It is easy to see why agreement has never been reached on a single dynamic or service-related oxidation stability tester.

EFFECTS OF WATER

Water may affect lubricating grease, and thus grease lubrication, in several distinct ways. The grease may change in consistency, becoming softer or occasionally firmer. The grease may change in texture, becoming, for example, less stringy and less adhesive. The water and grease may form an emulsion, which will usually have less lubrication capability than the original grease. Or the grease may form a soft emulsion and wash away. Various combinations of these behaviors have been experienced.

Most laboratories have designed nonstandard ways to measure this behavior. Some have mixed grease and water in a household-type mixer, then compared the dry grease with the wet mixture. Consistency, texture changes, and water rejection are generally observed.

More usual is the addition of water to grease in a worker cup or in the roll stability test chamber. Using the grease worker, the worker plate is sometimes modified; proportions of grease and water and number of strokes vary from laboratory to laboratory. If the roll stability tester is used, proportions of water and grease, temperature, and time of rolling vary. Observations are as before. In the worst case, the worked mixture of grease and water becomes completely fluid. In the best case, the grease is essentially unchanged, and water is rejected.

Water Washout

ASTM does have two tests which address the question of how grease behaves when subjected to a water stream or spray. In the older test, D1264, we try to measure how much grease will be washed out of a bearing under the conditions of test. A bearing is packed and mounted in a shielded, but not sealed, housing. The bearing is rotated at 600 rpm while water, at 100°F (38°C) or 175°F (79°C) and flowing at 5 cm^3/s, impinges on the housing.

Test results are affected by differences in grease texture and consistency. Firm, stringy grease will be pulled out of the bearing, the loss being reported as washout. Thus the test, although widely used, can give misleading results. Test precision, particularly with soft greases, is poor.

Water Spray-off

A recent ASTM test is D4049, which measures the amount of grease removed from a panel by an intense water spray. Grease is removed by a combination of two effects: solubility and impingement. Water-soluble greases will dissolve in the water deposited by the impinging spray. Most water-soluble greases, however, would not be subjected to this test.

During the test, a water spray hits the grease surface with considerable force. The blasting effect of the spray removes greases which do not have a great deal of cohesive strength. Such strength is imparted by suitable thickeners, oils, and additives. Soft greases and greases made with low viscosity base oils tend to do poorly in this test.

As to details of the test, grease is spread evenly on a horizontal metal plate to a thickness of 1/32 inch (0.8 mm). Water, at 100°F (38°C), is forced, at 40 psi (275 kPa), through a standard spray nozzle held 12 inches (30 cm) above the plate. Weight loss is determined after the panel has been sprayed for five minutes, then dried.

This procedure has been reported to correlate with behavior of greases in rolling mills in which intense water sprays are used.

CORROSION

As applied to lubrication, corrosion includes the deterioration of metal surfaces by chemical attack. This may be caused by some ingredient in the lubricant, or by external factors. Thus sulfur compounds in lubricants may corrode copper or lead, alkali in lubricants may attack aluminum, etc.

To test the effect of petroleum products on copper, ASTM D130 has long been used. This method was found to need minor modification when applied to lubricating greases. Thus a new method, D4048, has been standardized as a grease method.

Of more widespread interest, though, is that specialized form of corrosion commonly called rusting. When iron or steel is in contact with water and air, or aerated water, the metal may become rusted. If the water contains salt or acid, rusting is more rapid and more serious. When a grease-lubricated bearing surface comes in contact with water, that surface may become stained, then rusted (such staining can be the first stage of rusting). Rusting may be controlled, or even prevented, by incorporation into the lubricant of a suitable rust inhibitor.

To test for the rusting or rust-inhibiting characteristics of lubricating grease, we have ASTM D1743. This method has not been completely satisfactory and has been under continuing study. In its present form, a tapered roller bearing is packed with grease, then spun for one minute at 1750 rpm with a thrust load of 6 lbf (26.7 N). Most of the grease is thus removed from bearing surfaces; the rest is distributed in a thin layer. The assembly, without breaking contact between rollers and races, is dipped in distilled water and stored in a closed container for 48 hours at 125°F (52°C). After cleaning, the bearing cups are examined for evidence of corrosion. Modifications to this test were approved by ASTM in 1986 and should appear in the 1987 manual.

LOAD CARRYING TESTS

A lubricant separates bearing surfaces, permitting relative motion of those surfaces with minimal wear. If the amount of lubricant is not adequate, the lubricant film becomes so thin that only part of the surfaces remain separated. Some parts of the surfaces actually come into contact. Such contact, even if only sporadic, produces wear. At least, high spots--asperities--come into contact and wear or melt off (Figure 3.12).

Figure 3.12 - Function of an EP Agent

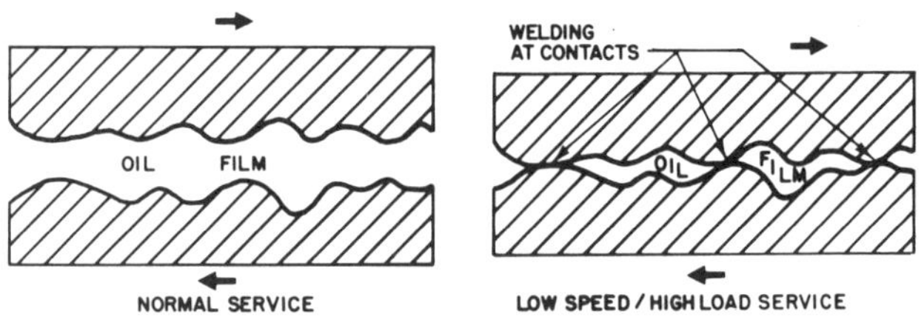

Illustration courtesy of Texaco Inc.

Heavy loads give the same result. The surfaces are pushed closer together, the lubricant film becomes very thin, and high points on the bearing surface carry some of the load. This condition is often known as **boundary** lubrication, and it is accompanied by a modest rate of wear of one or both bearing surfaces. Because of the ongoing wear, debris accumulates in the lubricant and, unless cleaned out, will gradually increase the wear rate.

With still more load, the surfaces may weld permanently, resulting in catastrophic failure, such as a broken bearing or shaft. This could be caused by roughness of the bearing surfaces as more and more metal is worn away, or it may be due to buildup of debris or from heat generated by the wearing away of metal. In any case, heavy loads cause lubricant films to become thin, leading to such bearing failure.

Lubricants differ in their ability to carry loads, sometimes by keeping the film thicker and sometimes by acting chemically on the surfaces, preventing them from welding. If the load is light or moderate, we can measure wear. If the load is heavy, wear may be so severe that we do better to seek physical evidence of surface damage or determine the load at which such damage occurs.

Three tests for load-carrying capability are widely used. One measures wear at relatively light loads; the others indicate more severe wear or welding.

Four-Ball Wear

In ASTM D2266 (Figure 3.13) we have a heavy, cuplike container in which three 1/2-inch (12.7-mm) hard steel balls are locked and coated with lubricant. A fourth ball is locked in a chuck and can be rotated while pressing against the three stationary balls. The rotating ball is thus rubbing against one area of each of the stationary balls.

The machine can operate at several speeds, temperatures, and loads and for different times. Maximum loading is 50 kgf (490 N). ASTM has standardized results for one set of conditions: 75°C (167°F), 1200 rpm, 40 kgf (392 N), and 60 minutes. At the end of the

Figure 3.13 - Four-Ball Wear Test

Photo courtesy of Southwest Div., Witco Corp.

test, the three balls are removed from the cup and wear scars on the three balls measured. The average scar diameter is reported as a measure of wear, low wear being indicated by a small scar.

Note that results are obtained only with hard steel rubbing against hard steel at these limited test conditions. Results would not be applicable with other metal combinations and at grossly different rubbing speed, temperature, and load. At this light load, seizure or welding does not take place, and we measure only wear. Correlation with service has never been established.

Four-Ball EP

This test, ASTM D2596 (Figure 3.14), has points of similarity to D2266 but is distinctly different. The machine is much more rugged, carrying as much as 800 kgf (7845 N). Test balls of the same size and metallurgy are held in the same configuration--three balls locked in a cup and covered with room-temperature lubricant, one ball held in a rotating chuck. The tester operates at one speed, 1770 rpm, and one load for ten seconds. The ball in the chuck is discarded. Test balls are taken from the cup and average scar diameter measured. The test is repeated at successively higher loads.

Figure 3.14 - Four-Ball EP Test

Photo courtesy of Southwest Div., Witco Corp.

Average scar diameters are corrected for Hertz indentation diameters in calculating corrected loads. These loads may be plotted in a diagram, forming a straight line. When a higher load causes deviation from this line, that point indicates incipient seizure. Tests continue at higher and higher loads. At some load, welding of the four balls occurs. The weld point is reported. Then a calculation, based on corrected diameters in the last ten runs below the weld point, is made and reported as load wear index.

Again, this test for the ability of a lubricating grease to carry heavy loads should not be overinterpreted. It concerns only hard steel rubbing against hard steel at one speed in essentially point contact, but actually small areas of contact. Validity of tests run at the highest loads is even more questionable, since the steel balls may be severely distorted at such pressures. If this should occur, the area of contact would be different from what might be expected geometrically from steel balls. Yet, this area is involved in the calculation of load wear index.

Results do not necessarily correlate with service but do seem to indicate low, medium, or high levels of load-carrying capability.

Timken Lubricant and Wear Tester

We have seen that two four-ball testers are used to evaluate load-carrying capacity of a lubricant. Another popular machine is the Timken Lubricant and Wear Tester, which can be used to measure wear at low loads but is more usually employed at higher loads to develop an estimate of load-carrying capacity.

In ASTM D2509 (Figure 3.15), the outer race (test cup) of a tapered roller bearing is mounted on a spindle rotating at 800 rpm. The cup rubs against an equally hard steel block which is mounted in a movable arm, at the end of which a weight is to be hung without shock. Geometry of the assembly causes the load application to be normal to the test cup. Grease is fed continuously to the area of contact, the machine is started, load is applied, and the test is run for ten minutes under full load. The load is then removed. Parts are cleaned and inspected.

Wear occurs all around the rotating cup but is concentrated on the block along the line of contact between cup and block. The scar is inspected. If it is smooth and shiny, there has been no failure of lubrication--or to say this differently, the lubricant has successfully carried this load, which is expressed as the weight applied at the end of the lever arm.

Successive runs are carried out at gradually higher loads until the scar produced is rough and irregular, indicating film rupture in the form of scoring or welding. Unfortunately, the onset of scoring is not

Figure 3.15 - Timken Test

Photo courtesy of Southwest Div., Witco Corp.

easily determined; thus extra runs must be made in an attempt to express the end point with more certainty. From these, we report the score value--the lowest load at which scoring does occur--and OK load--the highest load at which scoring is not observed.

The Timken test, like other EP tests, has poor precision. Results in the Timken test are not likely to agree with those in the four-ball test, since we are dealing with different speeds, loads, test duration, and geometry of contact. Results may not correlate with service but do indicate whether the product has EP characteristics and, broadly, levels of load-carrying capability.

OSCILLATING MOTION

Several grease tests involve oscillation. D3704 describes wear-preventive properties of lubricating greases when a ring and block similar to parts in a Timken tester are operated in oscillating motion. Wear is determined.

A second test, recently standardized, is D4170 (Figure 3.16), in which loaded ball bearings are oscillated, then examined for fretting wear. Fretting is a frequent problem in industry, since stationary machine elements are often subjected to vibration. If severe enough or prolonged, such vibration may cause indentation in bearing surfaces. Then, when the machine is started, the indentations cause rough running, leading to early failure.

Figure 3.16 - Fretting Wear Test

Photo courtesy of Southwest Div., Witco Corp.

Such fretting failures have been observed in wheel bearings of automotive equipment shipped a long distance before operation, in equipment stored for emergency service, and in equipment placed in position as construction of a plant begins, then left untouched through the construction period. Fretting failures are often reported as lubricant failure, which is not accurate, but a well-formulated lubricant can reduce the rate of fretting.

A third test, D3428, involves the effect of oscillating motion on loaded automotive ball joints. This complex test leads to reports of torque stability, wear, and brine sensitivity

ANALYTICAL TESTS

Analyses of lubricating greases tell about chemical composition but do not permit evaluation of grease properties or behavior. They will thus not be discussed here. Again, ASTM has many methods which are used for grease analyses. Additional methods may be found in other literature.

GENERAL

At the start of each ASTM book, we find a table of contents followed by a list of subjects. In this latter portion, there is a section on lubricants, within which is located a tabulation of tests on lubricating greases. All the ASTM methods discussed here and others, such as the analytical methods, are to be found there.

ASTM methods are continually changing. New methods are developed and added to the book. Old methods may be deleted or modified. It is thus wise to review test methods periodically. ASTM is a dynamic organization. When a test method is criticized or questioned, a committee is formed to investigate and seek improvement. If a change is approved, through a multilayered approval process, the method is revised.

Thus, although a test bears the same number as before, the current test may differ in some significant details from the method as it appeared in earlier editions.

Many ASTM methods have been coordinated with the (British) Institute of Petroleum, IP; the corresponding German standardization institute, DIN; and the American National Standards Institute, ANSI.

CHAPTER 4

CHARACTERISTICS OF TODAY'S GREASES

SINGLE PURPOSE GREASES

In the early days of this industry, our limited knowledge resulted in the formulation of greases each of which could be used in a small number of applications. Thus, for passenger cars, we had cup, chassis, wheel bearing, water pump, universal joint, hinge and latch, and other greases.

Consider the early motor car. Speeds were low, and so were operating temperatures. Calcium soap grease or aluminum soap grease, in different consistencies and made with different oils, could lubricate just about all parts of the vehicle. As roads and vehicles improved and more powerful engines became available, vehicles operated at higher speeds and loads. When applying brakes, wheel bearings rose in temperature above the capability of these products. For wheel bearings, sodium soap greases became the lubricants of choice. Sodium soap greases were also preferred as universal joint greases, because temperatures were high.

The sodium soap greases would wash out of chassis fittings. Thus calcium or aluminum soap greases or specially formulated soda soap greases were used in these parts.

Another problem arose. In case of severe water exposure, the water-soluble grease would wash out of the wheel bearings. This happened in driving rainstorms, when vehicles traversed flooded roads, etc. Thus, when multipurpose greases--both heat- and water-resistant--became available, they gradually displaced earlier products.

Industrial equipment went through the same evolutionary stages and was lubricated with similar products. Sodium or sodium-calcium soap greases lubricated electric motors; calcium (and some aluminum) soap greases were used for most general lubrication.

INCREASING VERSATILITY

Multipurpose and Specialty Greases

As more thickeners became available, better greases could be formulated for each application. This led to a proliferation of products. Simplification by use of multipurpose greases cut the number of products but did not eliminate the need for a variety of specialties. Some greases are used at high temperatures, others at low temperatures, and still others under a wide range of temperatures. Some are used under wet conditions, and some face difficult pumping conditions--long distances, narrow lines, and/or low temperatures. Loads may be heavy. Sometimes contaminants must be flushed out, requiring frequent relubrication.

Table 4.1 gives factors which must be considered in each of these types of service. Such factors are involved in grease formulation; they are important, too, for the user of greases.

Many more applications and requirements could be considered, but these cover the major points. In formulating a grease for special-purpose or multiservice applications, we must consider the range of operating conditions, suitable thickeners, acceptable fluids, beneficial additives, and appropriate tests. There is no such thing as a theoretically best grease, but there may be a best product for a particular application.

From Table 4.1, it becomes apparent that we should consider a few of the important characteristics of oils used in making greases.

TABLE 4.1

FORMULATION AND APPLICATION REQUIREMENTS

Service	Requirements
High temperature	High-temperature thickener, high dropping point Oil of high viscosity--petroleum or synthetic Oil of high flash point--low evaporation Firm consistency (preferred) Oxidation resistance
Low temperature	Low percentage of thickener Soft consistency Oil of low viscosity Oil of high viscosity index (preferred)

Service	Requirements
Low temperature (continued)	Oil of low pour point (at very low temperatures) Rust resistance (water may condense) Good low-temperature torque Good pumpability
Wide temperature range	High-temperature thickener Low percentage of thickener Good low-temperature torque Good pumpability Low evaporation Oil of low viscosity and high VI, preferably synthetic Oil of low pour point Oxidation resistance Rust resistance EP-antiwear often needed
Exposure to water	Low washout Low spray-off Firm consistency Water-resistant thickener High-viscosity oil Viscosity booster (helpful, not required)--polymer, wax, asphalt, etc. Rust resistance
Long pumping distances	Low apparent viscosity--good pumpability Good slumpability or feedability Low percentage of thickener Soft consistency Oil of low viscosity (preferred)
Heavy loads	Low wear test values High EP test values EP-antiwear additives Solid additives, if indicated Oil of high viscosity (preferred)
Contaminating environment	Frequent relubrication--good pumpability Low-cost formulation Firm consistency (for better sealing)
Multiservice	High-temperature thickener Oil of medium viscosity EP-antiwear usual Oxidation resistance Rust resistance Acceptable pumpability Water resistance

PETROLEUM OILS

Solvency of an oil for thickener has already been mentioned as an important characteristic. Several others are referred to directly or indirectly in Table 4.1.

Viscosity

Viscosity is the most significant property of a lubricating oil. It may be visualized as resistance to flow of a liquid. Technically, it is the coefficient resulting when shear stress is divided by shear rate. That coefficient of viscosity is constant for a normal oil operating at reasonable pressures and one temperature. As temperature rises, viscosity decreases; as temperature falls, viscosity increases. When we formulate low-temperature greases, better cold performance results from using a light (low-viscosity) oil.

Viscosity is the attribute which causes lubricant to be drawn into a bearing, forming a film which separates the surfaces. If the film is too thin to keep the surfaces apart, rapid wear will result.

As temperature rises, oil becomes "thin"--less viscous. If viscosity falls too low, the oil film will fail. Thus, for high-temperature greases, it is important to start with high-viscosity oil. The same consideration is involved for heavily loaded bearings. Under those loads, oil film becomes thin, and we are forced to depend on additives. It is best, therefore, to use oil of the highest viscosity, which will not interfere with pumping or dispensing the grease.

Viscosity Index

To represent the relative increases and decreases in viscosity as temperature falls and rises, an arbitrary scale has been developed. It is known as viscosity index (VI) and uses 100 to indicate viscosity changes expected of a paraffinic oil; viscosity changes expected of a naphthenic oil are set around 0. Some highly refined oils and synthetic oils have viscosity indices much higher than 100, indicating much less viscosity change with temperature than was visualized when the VI scale was created.

In terms of formulation and application of lubricants, it should be apparent that at temperature extremes, and especially where a wide temperature range will be faced, an oil of high viscosity index would be desirable.

Pour Point

When dealing with low-temperature service, pour point of the oil also must be considered. Pour point may be visualized as the temperature below which an oil will not flow under the force of gravity.

Low-VI oils thicken severely as temperature falls until high viscosity alone prevents the oil from flowing. This gives what is sometimes designated a "viscosity pour point."

High-VI paraffinic oils are apt to contain some dissolved wax. Part of this may be removed during refining, but a portion usually remains. As temperature falls, some of the wax crystallizes out, forming a gel or crystal network which prevents the oil from flowing. This condition is known as a "waxy pour point." A high-VI oil generally has a higher pour point than a low-VI oil of the same room temperature viscosity.

The thickener in a lubricating grease develops a much more rigid network than is formed by wax. Thus, for most greases used at usual ambient temperatures, wax is not a consideration. However, for some very low-temperature applications, the pour point must be considered. There have been some circumstances in which an oil of low VI and low viscosity pour point performed better than an oil of higher viscosity index, where wax apparently interfered.

For oils, we sometimes add pour point depressants. These are polymers which adhere to the wax and prevent it from crystallizing out until temperature has been brought lower. These are often ineffective in greases for at least two reasons. First, the thickener may interfere with the adherence of polymer to wax; second, soaps such as are used in soap-type greases are in themselves pour depressants. This behavior may not apply to all soap complexes and does not apply to nonsoap greases.

ADDITIVES

The variety of additives used in grease is extensive. Some are sold as chemicals, others as proprietary brand-name products. The literature on these materials is abundant. For detailed information, the usual reference books on greases are good sources.

APPLYING GREASE LUBRICANTS

High-Temperature Lubrication

Greases fail more rapidly as temperature of operation increases. The most obvious reason for failure lies in the melting point of the thickener or dropping point of the grease. The latter involves a complex of melting and bleed. Evaporation may be significant at high temperatures. Oxidation also increases rapidly as temperature rises. There are useful guidelines for heat resistance of greases in service which take all these factors into consideration.

Most mineral-oil-based greases (of adequate dropping point) will operate successfully to about 250°F (121°C). A smaller number can handle 300°F (149°C). A few mineral-oil-based greases can operate to about 350°F (177°C). Around this temperature, synthetic fluids are preferred or required. As service temperature rises, frequency of lubricant addition and relubrication must increase.

In industrial service, the following may be considered reasonable relubrication intervals for rolling element bearings (assuming eight work hours per day):

180°F (82°C), 6 months
220°F (104°C), 3 months
300°F (149°C), 1 month
380°F (193°C), 1 week
460°F (238°C), 1 day

These guidelines assume reasonable-size bearings operating at usual speeds and loads. If speed is high, bearing large, or load severe, relubrication intervals should be even shorter. Where service is severe and/or contamination is unavoidable, relubrication is best carried out with a centralized lubrication system, and lubrication intervals may be measured in hours or minutes.

For high-temperature service, greases must be of high quality. But quality is not a fundamental property of a lubricant. It is the result of many factors which, all together, lead to the performance sought. Test data, while indicating the capability of a grease to perform well in service, do not guarantee such behavior. That is learned only in actual operation in the field--in machinery, in vehicles, etc. This is the limitation of specifications and the reason that laboratory results must be confirmed in field tests.

Pumpability and Slumpability

In many applications, grease must be pumped--to reservoirs, to distribution systems, or to bearings. A grease gun is a simple pumping device. Ability of a grease to be pumped (pushed) is usually limited by the capability of the product to be drawn into the pump. This characteristic is often known as slumpability or feedability. There is no standard test for slumpability.

Greases made from high viscosity oils are poor in both slumpability and pumpability when temperatures are low. However, texture affects slumpability. Greases which are fibrous or stringy are generally good in feedability. Greases which are buttery--not fibrous--may not show up well at the suction side of a pump. But at the push side of a pump, the reverse is true, and the fibrous greases are poorer than nonfibrous products.

Compatibility

When greases made from different thickeners are mixed, the mixture may be poorer in service performance or physical properties than either of the component products. This lessening in performance capability is called incompatibility. It may show up in any of several areas, such as (1) lower heat resistance; (2) change in consistency, usually softening; or (3) decrease in shear stability. Mixtures which show none of these changes are considered compatible.

Incompatibility is not always caused by the thickener, since each of the greases in the mixture is a complete package--thickener, fluid, and additives. Sometimes the thickener of one grease is incompatible with the fluid or the additives present in the second formulation. If the mixture proves to be significantly softer, less shear-stable, or less heat-resistant than the original grease, the mixture must be deemed incompatible.

Incompatibility is best determined in service or in service-related tests; it is not predictable. Certain thickener combinations often have been found unsatisfactory and are generally so recognized. These would include lithium and sodium greases and organo-clay and most soap greases. Tests should be run on the specific greases of interest.

Consistency

Some discussion of consistency seems appropriate. First, the question of hard versus soft greases. Penetrations are determined at 77°F (25°C). If temperature falls, say to 32°F (0°C), the grease will be firmer by one or two NLGI consistency numbers. If temperature rises to around 110°F (43°C), the grease will be at least one consistency number softer than at 77°F (25°C). Grease which is stored at ambient temperature will feel soft in summer and hard in winter. It is likely, too, to be difficult to pump and high in torque in winter.

The same variation is found in different climates. In hot areas, NLGI Consistency Numbers 2 and 3 are usual; in cold locations, NLGI Consistency Numbers 1 and 0 are more common.

Where grease is handled in a dispensing system, the grade chosen may be related more to the capabilities of the system than to the requirements of the application. In service, consistency differences are important. Consider, for example, the use of grease in a gear case. The grease should be carried into the meshing gear teeth, which are thus kept covered with lubricant. If the grease is firm, the gear teeth may cut a channel through the grease and run dry, causing damage to the teeth. Unless operating temperature is high, therefore, soft-to-semifluid greases are ordinarily recommended.

In ball or roller bearings, carrying grease to the rolling elements is considered undesirable, since that grease will churn, soften, heat up, and work its way out through the seals. Thus a common recommendation for ball bearings is to have the bearing space only about one-third full of grease of firm consistency--often NLGI No. 2. If the bearings support a vertical shaft, the grease had best be still firmer--probably NLGI No. 3.

For a grease to perform well in service, its consistency must be chosen with regard to operating conditions, the mechanism to be lubricated, and the means of lubrication--hand, gun, system, or whatever. Consistency is the characteristic which often makes grease the lubricant of choice.

Greases may soften on working, then firm up on standing. This characteristic, thixotropy, is shown to different extents by many greases. Products of high thixotropy may become so firm in storage as to fall out of specification range. However, in terms of performance, the same behavior leads to lower torque when the bearing is turning and better sealing when the bearing is standing still.

TYPICAL GREASE PROPERTIES - SOAP THICKENED GREASES

In the following, we show properties of typical greases which are made from soaps, soap complexes, and nonsoap thickeners. Our object here is to describe the thickener-oil system without additives (unless the additive is part of the thickener system). Different manufacturers report varying values and properties for these systems. The values reported here are believed to be representative of commercial formulations.

When different products are compared, consistency, oil viscosity, and, if possible, thickener contents of the two greases should be as nearly identical as possible. Results reported may not always have been derived in this manner.

Aluminum Soap Greases

These products are smooth, clear, attractive gels, slightly stringy in texture, especially if made with high-viscosity oils. Their stringiness is often enhanced with additives. When heated, they tend to form heavy, stringy gels, which cause a sharp increase in power required to turn a bearing. This heavy gel formation varies with formulation but causes users to limit upper operating temperature to around 175°F (79°C), although dropping point exceeds 230°F (110°C).

At low temperatures, these products show more torque and are more difficult to pump than corresponding products made from other soaps.

Shear stability is poor. Oxidation stability, even without assistance from inhibitors, is excellent.

Water resistance is very good. Rust protection is good, again without help from inhibitors.

Cost is fairly high.

Sodium Soap Greases

These products, the most heat-resistant available in the early days of the grease industry, are sensitive to water. Some formulations will wash out even in fairly cold water, others in hot water. The soaps in these greases are present in the oil in fibers, usually quite long, thick, and tangled. Such fibers or fiber bundles are visible to the naked eye. Fibers are inherent--they are not added. When a sample of this long-fiber grease is drawn from a container, a long ropy portion is pulled out. By suitable changes in raw materials and processing, greases of medium and short fiber texture are produced. Sodium-calcium mixed base greases are also formulated to shorten and smooth the fibers of sodium soap greases.

Sodium greases have high dropping points--around 350°F (177°C)--but are not used much above 250°F (121°C) or occasionally 275°F (135°C), being limited by oxidation and bleed as well as softening.

Low-temperature pumpability and torque are adversely affected by the fibrous texture of the product.

Shear stability of most of these greases is satisfactory. Oxidation stability can be improved with additives.

While these products are poor in water resistance, in the presence of only a little water they form emulsions which protect metal surfaces from rusting by drawing the water away from the metal.

Cost is moderate, depending on the formula used. With tallow as the fat, cost is fairly low. When fatty acids and structure modifiers are included, cost rises.

Calcium Soap Greases (Hydrated)

These greases are smooth and buttery. When made with heavy oils, they become smooth and stringy. When heated, they lose their associated water, which may be water of hydration. They then become rough, bleed oil, and are not satisfactory lubricants. In service, they are thus limited to about 175°F (79°C), although dropping point is over 205°F (96°C). Soft grades may lose water and become unsatisfactory even below this temperature, but firm grades, being made with much more soap, have been used up to around 200°F (93°C).

At low temperatures, these greases are easy to handle.

Shear stability is fair to good. Oxidation stability is poor but can be improved with inhibitors. Since this grease is not used at high temperatures, the inhibited grease has excellent oxidation stability, especially during storage.

Water resistance is very good. Protection against rust is poor but can be enhanced with additives.

Cost is low, since this kind of grease is commonly made from tallow-type fats and fatty acids and hydrated lime--an inexpensive alkali.

Calcium Soap Greases (Anhydrous)

Of the available formulations of anhydrous calcium soap greases, the one made from 12-hydroxystearate soap is the most common. It is smooth and buttery. On heating, no water can be lost, but dropping point is not high--around 280°F (138°C)--and the product is limited to an upper temperature of about 230°F (110°C).

Low-temperature properties are satisfactory.

Shear stability is good. Oxidation resistance is acceptable and easily enhanced with additives.

Anhydrous calcium greases are very water-resistant but do not protect against rust unless suitable additives are incorporated.

These products are costly--comparable to lithium greases, which offer much better high-temperature properties.

Lithium 12-Hydroxystearate Greases

Although lithium stearate greases are still being made, the overwhelming majority of lithium greases produced today are derived from the 12-hydroxystearate soap. These products are smooth textured and stable to heating. Dropping points have been reported in a range from about 350°F to 400°F (177 to 204°C); most reported values are near the middle of that range. For long-term use, the upper temperature limit is around 275°F (135°C), somewhat better than we find with sodium soap greases. Even this "limit" can be exceeded if oxidation can be controlled.

At low temperatures, these greases are easily handled.

Under shear in laboratory equipment, these greases are excellent. It is again worth noting that shear rates in service are many times higher than we obtain in laboratory test equipment. This throws some doubt on the validity of extrapolating such laboratory test results to performance in service.

Oxidation resistance is acceptable and easily improved with antioxidants.

Water resistance is good, although not as great as with calcium or aluminum greases. Rust resistance must be furnished by additives.

Cost is fairly high.

COMPLEX SOAP GREASES

These greases are generally smooth, although some batches may show grain. They are stable to heating. Dropping points are generally over 500°F (260°C); the products may be used for short excursions to 350°F (177°C) if attention is paid to the tendency of most petroleum oil-based greases to oxidize rapidly above about 250°F (121°C). Therefore, relubrication at high temperatures must be frequent. At 350°F (177°C), relubrication every two weeks seems desirable (40 hours per week).

In all these products, oxidation stability, protection against rust, and load-carrying capacity can be provided or enhanced with additives. All have acceptable oil separation characteristics. Differences are significant, however, and will be discussed below.

Aluminum Complex Greases

Low-temperature behavior of these products must be rated as fair to good. At moderate temperatures, they are equivalent to other greases in pumpability.

Shear stability is good to excellent. Like many other greases, if thickener content is low, stability to working falls off significantly.

Water spray resistance is excellent. Water washout and some emulsion-type tests would rate these products as good.

In high-temperature, high-speed ball bearing life tests, such as ASTM D3336, these products are apt to give shorter lives than do most lithium complex or polyurea greases.

Cost is fairly high.

Calcium Complex Greases

This type of grease is found in many variations. Some types have high thickener contents. One ingredient of these products is calcium acetate, which provides EP properties. In high-thickener-content grades:

1. Load-carrying and antiwear qualities are high.
2. Water resistance is good to excellent.
3. Work stability is good.
4. Products tend to become firm in storage.
5. Cost is moderate.
6. Hardening may occur in storage or under pressure in lubrication systems.

If the grease is made with low thickener content:

1. Load carrying and antiwear need boosts from other additives.
2. Water resistance is fair to good.
3. Work stability is fair.
4. Product is fairly stable in storage.
5. Cost is fairly low.

Lithium Complex Greases

The easiest way to describe these products is to compare them with normal lithium 12-hydroxystearate greases. The two are quite similar, except in dropping point and in the heat resistance which this indicates. Dropping point of the complex is more than 100°F (56°C) higher than that of the soap grease.

They handle well at low temperatures.

Work stability, pumpability, and oil separation are good to excellent.

Bearing performance at high temperatures is very good to excellent.

Water resistance is good but may be deteriorated by some widely used additives. Some special formulations are excellent.

Cost is fairly high.

NONSOAP GREASES

Polyurea Greases

These smooth products are used at high temperatures; they are comparable to some of the complexes in this capability. Although used with all kinds of bearings, they have been particularly effective in lubricating ball bearings, such as are found in electric motors. This is indicated, too, in high-temperature bearing tests. Dropping point is generally slightly under 500°F (260°C), more like 470°F (243°C), but the products are sometimes usable to 350°F (177°C), as was noted with the complex-thickened greases.

These greases have oustanding resistance to oxidation. Their thickeners contain no soaps or other metal-containing constituents, which are, to varying degrees, pro-oxidants. The oxidation resistance of these products is attributed to the absence of metals. However, their rust-preventive and load-carrying additives do contain metals. As another possibility, one or more of the reactants used in making polyurea thickener could be an antioxidant, making the final thickener resistant to oxidation.

Handling at low temperatures is satisfactory.

Work stability as measured in laboratory tests is quite poor. As has been mentioned earlier, this behavior should not be extrapolated to service, because shear rates in service are many times higher than is usual in the laboratory. However, the degree of shear sensitivity shown is so high as to raise the possibility that polyurea greases intended to be used with ball bearings, where shear rates are generally low, and roller bearings, where shear rates are generally high, should have different formulations.

Water resistance is satisfactory--in some grades, excellent.

Rust resistance requires the use of special and effective rust inhibitors.

Cost is fairly high.

Organo-Clay Greases

These smooth-textured greases have outstanding heat resistance, since the thickener will not melt, at least up to the temperature at which its constituent oil evaporates, flashes off, or burns. However, since petroleum oil is the limitation, dropping point is given as over 500°F (260°C) and maximum use temperature as 350°F (177°C). These are the same as was reported for the other high-temperature greases of the complex and nonsoap series.

All these high-temperature greases can be used for occasional and short excursions to higher temperatures than are usually recommended. In each case, the need for frequent relubrication remains. For example, since this thickener has no melting point, organo-clay greases have been used in applications in which short-term temperatures could hit 500°F (260°C); relubrication is then required after only a few hours of this high-temperature service.

Low-temperature properties are satisfactory. However, many organo-clay greases are formulated for high-temperature applications in which oil viscosity is high. Such formulations, of course, will have poor low-temperature properties.

Work stability must be rated as fair to good.

Oxidation stability and rust resistance are satisfactory when enhanced with additives.

Water resistance is excellent.

Cost is fairly high.

CHAPTER 5

WHEN TO LUBRICATE WITH GREASE

In lubricating any given mechanism, a fundamental choice must be made. Will the lubricant be a liquid--oil, or a solid, or an intermediate product--grease--which may range from semifluid to semisolid? The choice is made, in fact, by the machine designer, who looks at expected operating conditions and the properties of available lubricants and decides on the lubrication system to be used.

The designer finds several advantages in using greases. One is cost. Greases assist seals; oils require more effective, more expensive seals. If vertical shafts are involved, effective sealing of oil is quite difficult and costly. Sealing greases is less complex and more economical.

Another significant cost saving is due to the decreased complexity and space required when grease is used. To lubricate a bearing with oil, we frequently find an oil-filled sump below the bearing and a pumping device which lifts oil from the sump to the bearing. The oil then drains back to the sump. Compare this with the corresponding grease-lubricated bearing. Neither pump nor sump is required. A housing around the bearing holds a small supply of grease, which does not drain away. Such a device, which is less complex and more compact, will be less expensive than an oil-lubricated bearing.

Many other characteristics of greases are helpful to both designer and user of the equipment. Some of the advantages in using grease are highlighted on the following pages.

TO DECREASE DRIPPING AND SPATTERING OF LUBRICANT

Greases assist seals. Thus, grease tends to stay in its housing and not leak or drip out. This is particularly true as seals become worn. The effect on housekeeping is obvious. In processing foods, pharmaceuticals, or other products for which contamination must be avoided, grease minimizes problems.

TO DECREASE FREQUENCY OF LUBRICATION

Machine design frequently requires the placement of bearings where they cannot be easily serviced without shutting down production. In these applications, greases can provide lubrication for long periods of time with only infrequent lubrication. Electric motors in plants are often lubricated annually or even less often. Some electric motors and gears are lubricated for life at the factory. Obviously, infrequent lubrication requires high-quality greases.

TO SEAL OUT CONTAMINANTS

Since greases make seals more effective, they assist in keeping out contaminants. In mining operations and steel mills, dust, dirt, and water must be excluded. In paper mills, corrosive water and pulp are potential contaminants. In chemical plants and many other environments, corrosive gases and liquids must be kept out.

If corrosive contaminants do enter the bearing cavity, grease is more protective of the metal surface than oil would be. When a machine is stopped, oil drains away from bearing surfaces, leaving them essentially unprotected. Grease does not drain away from those surfaces. Thus, a protective grease film remains on the bearing surfaces to protect against corrosion.

FOR INTERMITTENT OPERATION

The fact that oil drains away when a machine is stopped, but grease does not, is significant also whenever equipment operates intermittently, changes speed, or reverses direction. On start-up, bearings are dry until oil has been drawn in between surfaces, forming a lubricating film. Grease does not drain away when the device is stationary; on start-up, a lubricating film forms immediately. Thus, for intermittent operation, grease protects against wear that would be found using oil as the lubricant.

TO SUSPEND SOLID ADDITIVES

Some solids have lubricating properties in varying degrees. When mixed with oils or greases, the mixtures often perform better than the oil or grease alone does. Typical solids are graphite, molybdenum disulfide, and zinc oxide.

When mixed with oil, the solids tend to settle out, leaving very little, if any, in suspension. The solids can be ground very fine at considerable cost; yet the amount remaining in suspension is small. When mixed with grease, even coarse solids remain in suspension indefinitely at any proportion desired.

WHEN EXTREME OPERATING CONDITIONS EXIST

Under very severe or extreme operating conditions, greases are quite effective lubricants. Several types of extreme conditions are described here.

High Temperature

Many industrial operations subject equipment and lubricants to very high operating temperatures. Kiln car, ladle car, and furnace door bearings are examples. Oil lubrication tends to develop too thin a film and poses a significant fire hazard. High-temperature greases, sometimes containing solids, give quite satisfactory lubrication.

High Pressure

Under heavy loads, films in oil-lubricated bearings are so thin that bearing service life tends to be short. The slightest particulate contamination can cause the film to rupture,accelerating wear. Load-carrying or EP additives, of course, assist the lubricant in handling heavy loads. High loads are common in industrial service and are found also in automotive equipment.

Shock Loading

Shock loading is an instantaneous and severe increase in stress. Such loading tends to rupture a thin lubricant film, leading to rapid wear. Greases appear to cushion against such sharp rupturing of the film. Shock loading occurs, for example, in rolling mills when a thick slab hits rollers which have been preset to a lesser thickness.

Low Speed Combined With High Pressure

Grease is usually the preferred lubricant when bearings are operating at speeds below 100 rpm at high load. If a very high-viscosity oil is used, an adequate lubricating film will be formed. However, such an oil has very poor pumpability and often cannot be employed. A lower-viscosity oil will not form an adequate lubricant film under these conditions.

Because of their high apparent viscosity at low speeds, greases will form a lubricant film of adequate thickness. Such a film, despite the high load, will lubricate successfully under these adverse circumstances.

WHEN MACHINE PARTS ARE BADLY WORN

A mechanism designed for oil probably will be lubricated with oil for the life of the machine. However, as equipment grows older, wear causes machine clearances to increase until adequate oil films can no longer be maintained in service. At this juncture, a change to grease is helpful, since with increased clearance, shear rate of the moving mechanism--bearing, gear, or whatever--decreases. At low shear rate, grease films are much thicker than oil films, providing adequate lubrication.

A change to lubricating grease permits the machine to continue in service at least until a maintenance shutdown can be scheduled. Some gear cases designed for use with gear oil have operated for years with greases ranging from NLGI No. 000 to No. 2 in consistency.

WHEN NOISE REDUCTION IS IMPORTANT

As machines become worn, gears, bearings, and shafts develop looseness. Loose, sloppy fits lead to noisy operation. Under such circumstances, the thick films provided by greases cushion the excessive movement and significantly reduce the noise level.

Most lubrication is carried out with oils. We have tried to show here that under many situations encountered in service, the properties of greases are superior, making them the lubricants of choice.

CHAPTER 6

WHICH GREASE TO USE WHEN

Grease marketing is a balancing act among four interested groups. The grease manufacturer wants to make as few products as possible and place them in as few packages as will be acceptable. The grease marketer, sometimes but not always representing the same company, also prefers to handle and distribute as few products as possible in as few packages as are required. These interests are similar but may not be identical.

The third interest is the equipment builder. Once his equipment goes into service, he wants it to operate to the user's satisfaction for a long time. He wants suitable lubricants for his machines, regardless of whether such products are widely available or must be developed specially.

The fourth interest is on the part of the user of lubricants who must operate the machines in the plant. The plant may be large and have many machines requiring a variety of lubricants in different quantities and different package sizes. Someone in the plant is brought in or trained to lubricate all the equipment on a reasonable schedule. He usually seeks to do this with the smallest number of products--if possible, only one.

Unfortunately, there is no all-purpose grease. But, by considering the properties already discussed, we may come to a multipurpose or multiservice grease. It must operate at the highest temperature encountered in service, yet be pumpable at the lowest ambient temperature. It must be stable to oxidation and prevent rusting. It must not wash away if water is present. It must handle the variety of speeds and loads-oh, yes, and do all these things economically.

Most grease users develop their own expertise. They find it helpful to work with knowledgeable suppliers, who make available lubrication engineering service and often support their marketing with informative booklets. This approach is recommended, since suppliers may be familiar with each specific machine and its operating requirements. Most users find their major needs met with one or more multi-service greases and minor requirements with a variety of specialties.

The laboratories of grease suppliers use many tests; most but not all of them are ASTM tests, which indicate levels of expected performance. Laboratory developments are often supported with field testing. Since suppliers do not all use identical tests, conclusions as to the "best" grease for a given service will vary. In addition, companies have patent positions, supply positions, or specific expertise which makes them favor one grease over another. Thus many different multi-service and special-purpose greases are being marketed.

GREASE APPLICATION GUIDE

The grease application guide shown in Table 6.1 will assist the end user in making a preliminary selection. Thickener is given so the user can be aware of the possibility of incompatibility among products and take suitable precautions. Properties are shown for greases made from ten thickeners along with remarks and an indication of important uses. Some of the ratings given are subjective and will vary significantly from one supplier to another. But it is believed that user and equipment builder can find a helpful starting point in this table. The greases described are fully formulated products, as found in the marketplace.

To illustrate marketing situations, four end-use markets have been selected. One is the automotive aftermarket. Three are industrial. One of these is a high-volume market--primary metals. Two have unique operating requirements: food processing and textiles. Builders' recommendations are given strong consideration but are not followed if circumstances favor alternative products.

AUTOMOTIVE AFTERMARKET

Motor vehicles usually use a grease of NLGI No. 2 consistency number (softer in cold weather) in most chassis, power train, and wheel bearing applications. Widely used greases can withstand high temperatures, resist breakdown due to mechanical action, resist

GREASE APPLICATION GUIDE

Properties	Aluminum	Sodium	Calcium-- Conventional	Calcium-- Anhydrous	Lithium	Aluminum Complex	Calcium Complex	Lithium Complex	Polyurea	Organo-Clay
Dropping point (°F)	230	325-350	205-220	275-290	350-400	500+	500+	500+	470	500 +
Dropping point (°C)	110	163-177	96-104	135-143	177-204	260+	260+	260+	243	260+
Maximum usable temperature (°F)	175	250	200	230	275	350	350	350	350	350
Maximum usable temperature (°C)	79	121	93	110	135	177	177	177	177	177
Water resistance	Good to excellent	Poor to fair	Good to excellent	Excellent	Good	Good to excellent	Fair to excellent	Good to excellent	Good to excellent	Fair to excellent
Work stability	Poor	Fair	Fair to good	Good to excellent	Good to excellent	Good to excellent	Fair to good	Good to excellent	Poor to good	Fair to good
Oxidation stability	Excellent	Poor to good	Poor to excellent	Fair to excellent	Fair to excellent	Fair to excellent	Poor to good	Fair to excellent	Good to excellent	Good
Protection against rust	Good to excellent	Good to excellent	Poor to excellent	Poor to excellent	Poor to excellent	Good to excellent	Fair to excellent	Fair to excellent	Fair to excellent	Poor to excellent
Pumpability (in centralized systems)	Poor	Poor to fair	Good to excellent	Fair to excellent	Fair to excellent	Fair to good	Poor to fair	Good to excellent	Good to excellent	Good
Oil separation	Good	Fair to good	Poor to good	Good	Good to excellent	Good to excellent	Good to excellent	Good to excellent	Good to excellent	Good to excellent
Appearance	Smooth and clear	Smooth to fibrous	Smooth and buttery	Smooth and buttery	Smooth and buttery	Smooth and buttery	Smooth and buttery	Smooth and buttery	Smooth and buttery	Smooth and buttery
Other properties		Adhesive & cohesive	EP grades available	EP grades available	EP grades available, reversible	EP grades available, reversible	EP and antiwear inherent	EP grades available	EP grades available	
Production volume and trend[1]	No change	Declining	Declining	No change	The leader	Increasing	Declining	Increasing	No change	Declining
Principal uses[2]	Thread lubricants	Rolling contact bearings	General uses for economy	Military multiservice	Multiservice automotive & industrial	Multiservice industrial	Multiservice automotive & industrial	Multiservice automotive & industrial	Multiservice automotive & industrial	High temp. (frequent relube)

Notes

[1] Lithium grease over 50 percent of production and all others below 10 percent.

[2] Multiservice includes rolling contact bearings, plain bearings, and others.

washing out, are stable to oxidation, protect against fretting wear or rusting, are easy to handle, and satisfactorily lubricate all mechanical parts of the vehicle. Exposure to high temperature is particularly severe in the case of front wheel bearings of cars equipped with disc brakes. For this application, greases indicated in Table 6.1 as having a maximum usable temperature of 350°F (177°C) should be selected. Other properties shown in that table can be compared with the needs of the application. In heavily loaded trucks, high load-carrying capability is also needed.

We can obtain some guidance from recommendations of the builders. For wheel bearings, one builder uses a lithium grease containing polyethylene and molybdenum disulfide. Another prefers lithium complex grease.

Working groups in ASTM, NLGI, and SAE* are developing an automotive grease classification system in which chassis and wheel bearing greases having two levels of performance are described. The products are delineated in terms of performance tests instead of their chemical composition.

PRIMARY METALS--STEEL MILLS

A major integrated steel producer uses at least 15 greases--considerably more if we include all grades of the various lubricants. Companies of that size specify, either in composition or performance, and sometimes both, the properties of products they are willing to purchase. The bulk of their usage consists of EP multipurpose greases operating at temperatures below 135°C (275°F). Their greases are often pumped long distances, then applied through complex dispensing systems. Water may be sprayed in the vicinity of the bearings in great volume--either for scale removal or for cooling, or both.

From Table 6.1, it is easy to select any of a number of products which can meet these requirements. Particular attention is paid to pumpability, water resistance, and load-carrying capability. Cost is also important, since the volume of grease used in steel mills is high. Some mills have separate lubrication systems feeding bearings operating at lower temperatures. These may use calcium greases, largely as a cost-saving measure.

The greases most commonly used in steel mills appear to be lithium and aluminum complex greases.

*Society of Automotive Engineers, 400 Commonwealth Drive, Warrendale, PA 15086.

FOOD PROCESSING

Greases are used for machinery in a variety of food processing plants, such as bakeries, breweries, dairies, fruit and vegetable packagers, soft drink canners, and, incidentally, can-making plants. Grease is used to lubricate some of the machinery in these plants.

The U.S. Department of Agriculture conducts inspection and grading programs covering, among others, meat, poultry, and rabbit processing plants. Lubricants are among the "nonfood compounds" used in a federally inspected plant, but not expected to become components of the food. On application, suitable lubricants may be authorized by the Food Ingredient Assessment Division (FIAD) for use in such plants. Authorized products are listed in Misc. Publication No. 1419. The same list is used for the Fishery Products Inspection Program of the Department of Commerce.

Lubricants identified in Publication 1419 as Code H1 are approved for incidental contact "on equipment and machine parts in locations in which there is exposure of the lubricated part to edible products." Lubricants identified as Code H2 may be used "where there is no possibility of the lubricant or lubricated part contacting edible products." In no case is a lubricant approved as an ingredient of the food; that would make it a food additive, which falls under different regulations.

H2 lubricants are regular commercial products (containing no poisonous ingredients).

H1 lubricants are made from a limited number of ingredients which are "generally recognized as safe" (GRAS) and/or appear in another USDA publication, AH562, under "Lubricants." In this list, we find the following types of thickeners:

- Aluminum stearate
- Aluminum complex (aluminum stearoyl benzoyl hydroxide).
- Organo-clay (dialkyldimethylammonium aluminum silicate. . .).
- Polyurea.

Acceptable additives are also spelled out. Each must be used at the minimum amount which would attain the technical objective sought with that additive. The oil used must be pure as is defined in the Code of Federal Regulation (CFR) under Title 21, Part 178, Section 178.3620(a) or (b). The oil is specified by color and ultraviolet absorbance.

The same CFR Title 21 publishes regulations pertaining to the Food and Drug Administration (FDA). Here, too, regulations related to lubricants with incidental food contact are published. In paragraph 178.3570, a list appears which agrees substantially with the USDA list. The FDA GRAS list does not include aluminum stearate as one of the acceptable grease thickeners.

Commercial products meeting H1 requirements are often marketed as "food industry" greases or "food machinery" greases. The word "food" is not used in the names of H2 greases, since they are standard commercial products.

TEXTILES

In the textile industry, multipurpose greases are utilized. However, because of the particular conditions in this industry, special modifications of such products are preferred. Thus, to minimize staining, light-colored greases are used, sometimes including a requirement for scourability. Humidity in textile plants is kept high to cut down the development of static charges. Rust-preventive properties are thus important.

Textile greases are also required to have load-carrying capabilities, be resistant to oxidation, and have good adhesion. This last requirement applies particularly to loom lubricant, which is a semifluid grease of about NLGI 000 consistency. Loom lubricant must also have low resistance to flow or develop low torque. Since one textile plant may contain hundreds or thousands of looms, a lubricant resistant to flow or contributing high torque would require a great increase in power consumption and could develop enough drag to cause breakage of thread.

In some textile operations, as in dryers, temperatures are high, leading to a preference for greases of high dropping point.

From the data about products in Table 6.1 and the preceding discussions about the grease lubricants needed in different industries, we see that marketing greases is not simple. It requires knowledge of products and their characteristics and limitations; of industries and their magnitude, their equipment, and the requirements of that equipment; and of the people who purchase lubricants and run the equipment. When all these desires and needs are matched, the marketing programs of the grease industry are bound to succeed.

CHAPTER 7

TROUBLESHOOTING GUIDE

This chapter provides a simplified approach to solving grease application problems. Although the format follows a component application approach, the latter part deals with environmental conditions--moisture and temperature. Therefore, when problems are encountered, the guideline should be reviewed in its entirety to ensure that all aspects of the situation are covered.

GREASE TROUBLESHOOTING

Application	Symptom	Possible Cause	Check For
Bearings (Assumes that correct bearings are in service.)			
Rolling contact	Noise	Condition of bearing	Worn or brinelled bearing.
	High bearing temperature	Overgreasing	Too frequent application. Bearing packed too full. Excessive grease charged per servicing.
		Starvation	Insufficient application frequen
		Incorrect product	Incorrect base oil viscosity. Deficient load-carrying ability (EP quality).
	Excessive leakage	Seals	Mechanical damage. Incorrect installation.
		Overgreasing	Too frequent application and excessive amount applied.
		Incorrect product	Grease too soft for application or softening in service.
		Incompatibility of grease	Admixture of greases.
	Frequent bearing replacement	Excessive wear	Load-carrying ability (EP of grease to handle shock loadir Starvation. Contamination, dirt, and rust. Normal bearing life exceeded. Installation malpractice. Grease too stiff, causing chann
		High temperature	High operating temperature.
		Misalignment	Correct alignment.
Plain type	Overheating	Improper distribution in bearing	Grease too stiff. Incorrect grooving.
		Starvation	Infrequent application. Defective/plugged lubricator.
		Incorrect grease	Mechanical stability of grease i service.
	Excessive wear	Starvation	Infrequent application Defective/plugged lubricator.
		Incorrect grease	Inadequate load-carrying ability of grease. Temperature range of grease.

Application	Symptom	Possible Cause	Check For
Gears			
Enclosed	Excessive leakage	Grease too soft for application Incompatibility of greases	Product penetration. Milling down of product. Admixture of greases.
	Noisy gear box	Lack of lubrication	Improper lubricant level. Grease too stiff.
	Overheating	Lack of lubrication	Improper lubricant level. Grease too stiff.
		Churning	High grease level. Grease too stiff.
	Tooth breakage	Not usually lubricant-related	
	Pitting	Mostly improper design and fatigue-related	While not generally lubricant-related, a heavier grease or base oil may retard progression of pitting.
	Wear and scoring	Lack of lubrication	Improper lubricant level.
		Incorrect product selection	Consistency, EP quality, and base oil viscosity.
		Abrasive wear	Lubricant contamination.
		Alignment	Not lubricant-related.
Open	Gear wear	Lack of lubrication	Incorrect lubricant. Incorrect application frequency.
	Buildup on gears or in roots	Excessive lubricant	Frequency of application. Proper type of lubricant. Airborne dirt.
Sliding surfaces	Nonuniform motion (stick-slip)	Lack of lubrication	Frequency of application. Proper type of grease--EP qualities or adhesiveness.
Universal joints	Excessive wear	Insufficient lubrication	Lubricant EP qualities. Lubricant high-temperature qualities. Application frequency. Slumpability of grease.
Electric motors	Electric malfunction High temperatures	Excessive grease leakage	Lubrication frequency and quantity applied.
Couplings	Dry coupling	Excessive grease leakage	Damaged seals. Consistency of grease. Keyway openings. Initial fill.
	Hardened grease	Centrifugal separation	Proper grease quality.
	Excessive wear	Incorrect grease	EP qualities of product.

Application	Symptom	Possible Cause	Check For
Centralized lubricators	No grease to points of application	Depleted reservoir	Fill with proper lubricant.
		Pump malfunction	Air/electrical supply.
		Plugged metering blocks	Plugging and proper grease.
		Airbound system	Bleed as required.
		Damaged feeder lines	Inspect and correct.
	High system pressure	Plugged metering devices	Check and clean.
		Malfunctioning relief valve	Check and repair.
		Grease consistency too hard	Product recommendation.
Wet applications	Noise--high wear	Lack of lubrication	Application frequency. Type of grease in service.
		Washout of lubrican	Extended application frequenc Grease consistency. Incorrect thickener type.
	Excessive corrosive condition	Incorrect lubricant selection	Product's inability to absorb wa Inability to maintain structure. Rust inhibitor additives.
High temperature	Noise--high wear	Lack of lubrication	Application frequency. Type of grease in service.
	Excessive leakage	Improper grease Incompatability of greases	Thickener type. Base oil viscosity. Consistency of grease. Admixture of greases
		Seals	Not lubricant-related (unless grease and seal are incompati
	Grease hardening	Improper grease	Oxidation stability of grease. Thickener type.
		Infrequent relubrication	Frequency of relubrication.
Low temperature	Component motion restricted	Incorrect grease	Grease with low torque quality Base oil viscosity.
	Difficult application	Incorrect grease	Pumpability qualities. Base oil viscosity. Consistency.
	Freeze-up	Water in system	Water contamination. Lubricant's ability to absorb/s water.

CHAPTER 8

PACKAGING, STORING, HANDLING, AND DISPENSING GREASE

GREASE PACKAGES

The NLGI reports that in 1985, 430 720 000 pounds (195 370 000 kg) of grease was shipped in the United States. A small proportion of product was shipped in bulk and the rest in packages, as follows:

TABLE 8.1

GREASE PACKAGES (AND BULK)

	Percent	Millions of Pounds
Bulk	14.5	62.5
Packages		
Jumbo bins	1.0	4.2
Drums	29.2	125.7
Kegs	16.2	70.0
Pails	15.8	68.0
Cartridges	21.2	91.2
Other small packages	2.1	9.2
Total	100.0	430.7

Cartridges

Although several cartridge sizes appear on the market, the principal size contains 14 ounces (0.40) of product. If we take this as the average cartridge, approximately 104 000 000 cartridges were used to package greases in 1985. Cartridges are used almost exclusively in loading portable grease guns.

The bodies of these tubes generally have been made of fiber with an oilproof liner. One end is built in by the cartridge manufacturer (Figure 8.1). Each tube is filled through the open end, which is then closed with a metal or plastic slip cover. Some manufacturers have gone to an all-plastic cartridge--both body and ends (Figure 8.2).

Figure 8.1 - Typical Cartridges

Photo courtesy of Southwest Div., Witco Corp.

Figure 8.2 - Plastic Cartridges

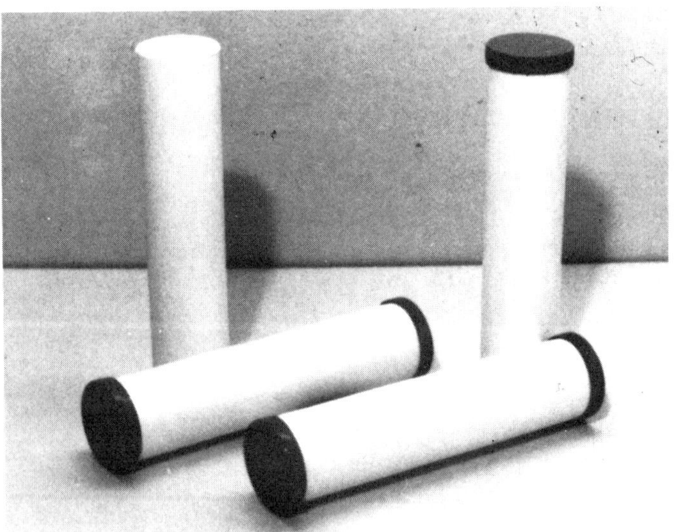

Photo courtesy of Southwest Div., Witco Corp.

Grease in cartridges is more expensive than it would be in a larger container. However, in many applications, the convenience of carrying one or more cartridges for rapid reloading of a grease gun is much more important than the cost involved.

Cartridges are used in large quantities in construction, farming, and underground mining. Most cartridges are filled with multipurpose greases, minimizing the possibility of misapplying the product.

Some grease guns will not work with cartridges. Thus, the user must be sure that the gun and cartridge are matched--in both size and configuration.

Other Small Packages

At one time, many greases were packaged in small containers. One plant's inventory would include 1-, 5-, and 10-pound (0.45-, 2.3-, and 4.5-kg) cans. Over the years, simplification and consolidation caused the 10-pound can to be eliminated and the other sizes to decrease in availability. Consumers still found a need for a small amount of grease and sometimes took it out of a cartridge.

More recently, plastic tubs have come into use, usually in 1- and 5-pound sizes (Figure 8.3). Tubs have replaced many of the metal cans. (For the uninitiated, since small cans have been hard to get, a number of producers have been marketing greases in plastic tubs similar to some used with foods.)

Figure 8.3 - Plastic Tubs

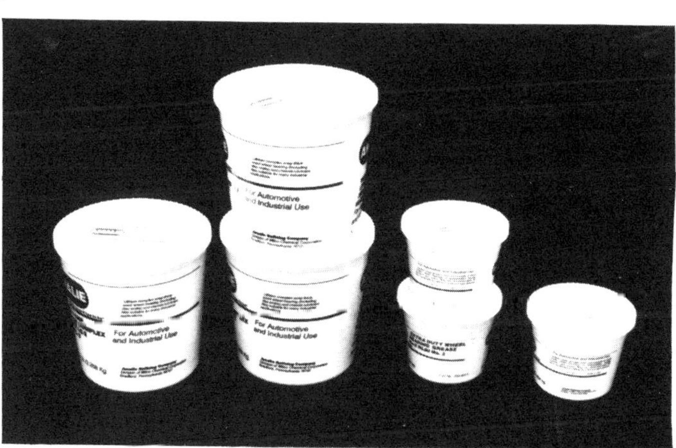

Photo courtesy of Southwest Div., Witco Corp.

Small squeeze tubes are also used to package greases for low-volume, specialized applications or as convenience packages for use around the shop or home. Typical tube capacities range from 1/2 ounce (14 g) to 8 ounces (227 g).

Other small packages sometimes used to hold lubricating greases are jars and ointment tins.

Pails

When grease containers were standardized, the 35-pound (15.9-kg) pail was chosen. This has the same body as the 5-gallon (18.9-dm^3) pail, but a head which is completely removable. The Canadian pail holds 6 U.S. gallons. The steel pail cover has a number of tabs or "lugs," which are crimped down after filling, making a tight package.

Plastic pails have been replacing steel pails (Figure 8.4). The change is due to economics--cost of the pail and a weight advantage which reduces shipping charges. Filled plastic pails should not be stacked too high, or the stack may buckle. Steel pails of standard gauge are more rigid.

Figure 8.4 - Pails

Photo courtesy of Southwest Div., Witco Corp.

Polyethylene plastic pails become brittle in very cold environments. They also become soft if the temperature of grease during filling is too hot or if storage temperature is high. Incompatibility with some components of greases may also present problems.

Kegs

Some years ago, a "quarter drum" containing 100 pounds (45 kg) of grease was widely used. In a standardization effort, the 120-pound (54-kg) keg with full-opening lug cover was selected. Over time, lighter gauge steels became popular. A composite container using a plastic liner and a fiber shell has also been used successfully (Figure 8.5).

Figure 8.5 - Kegs

Photo courtesy of Southwest Div., Witco Corp.

Drums

More grease is packaged in the 400-pound (181-kg) drum than in any other package. This drum is a 55-gallon (208-dm^3) full open-head container. After filling, a gasket, a steel cover, and a locking ring are applied to seal the container (Figure 8.6).

The grease pail, keg, and drum can be supplied with 2-inch (5-cm) openings through which a pump can be inserted without removing

Figure 8.6 - Drum

Photo courtesy of Southwest Div., Witco Corp.

the cover. This capability is a convenience and also decreases the possibility of contamination.

To ensure almost complete removal of grease from pails, kegs, and drums, follower plates can be provided. In such cases, a center opening in the lid is generally required (Figure 8.7).

Of the grease containers, only the full drum is sometimes reconditioned for reuse. As steel drums have become more expensive, economics have favored drum reconditioning. However, there are compensating disadvantages: Cost of handling and shipping the empty drums is increasing, and disposing of waste grease and cleaning solutions becomes ever more difficult and costly.

Experiments are going on in an attempt to develop an economical alternative to the steel drum. Fiber and plastic have been suggested. Such drums would be single trippers, requiring disposing of both containers and waste grease in a legal manner.

Drums are often moved with a two-wheel hand truck, sometimes with a forklift truck. Where larger lift trucks are available, drums are palletized and handled as a unit. To lower the cost inherent in handling a large number of drums, some consumers have engineered their plants to receive grease in still larger containers, often called jumbo bins.

Figure 8.7 - Pail with follower plate

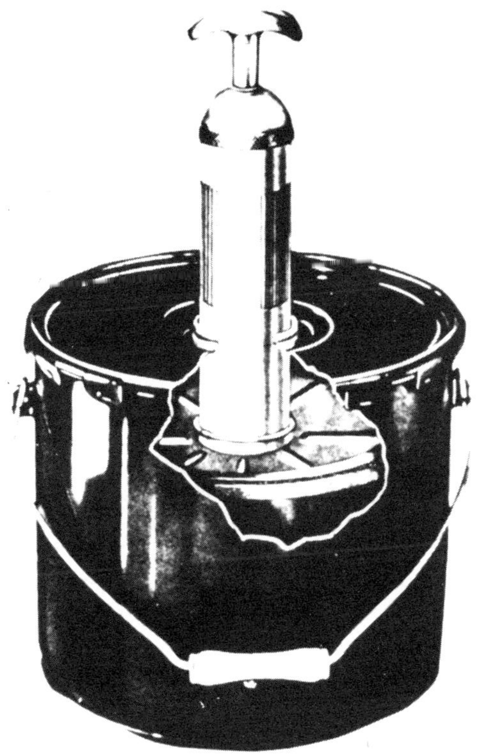

Photo courtesy of MSI Ltd.

Jumbo Bins

Although these have not been standardized, most bins hold 3 000 pounds (1360 kg) or 4 000 pounds (1814 kg). Bins are transported on trucks, sometimes of special design, to the customer's plant. There the bins may be lifted off by forklift or crane, the empty bin being returned for refill. In some cases, grease is pumped from the bin, still on the truck, to a customer's tank. Each such transfer eliminates the handling of eight or ten drums.

Bins are sometimes left on site at the customer location, and refilled from subsequent bulk shipments.

Bulk Delivery

Many years ago, semifluid greases were delivered to mines in tank trucks and stored above ground in tanks. From there, the grease was moved underground by techniques which varied from company to company.

When multipurpose greases became available, large grease users, particularly steel companies, started investigating purchasing such products in tank trucks. The mill first had to install large tanks in the mill and costly systems for distributing that grease throughout the mill. The grease supplier faced the problem of using a tank truck or a specially designed vehicle for transporting grease from grease plant to steel plant.

If a tank truck was used, a great deal of product was left in the truck after each delivery, increasing delivery costs beyond reason. Some trucks were scraped out with hoes to cut down the amount of grease returned as scrap. Some trucks were driven up on ramps, again to increase grease delivery. These moves were not effective.

Specially designed tankers were built--some as a single chamber for large deliveries, such as 30 000 pounds, (13 600 kg) and others with three chambers or three individual tanks. These were sometimes fitted with follower plates to improve delivery.

Follower plates of large diameter can easily become tipped or cocked while filling. Then some of the grease would flow to the back side of the plate, decreasing the possible delivery and requiring cleaning and adjusting before the next trip. Ingenious mechanisms were created to eliminate or at least minimize this possibility.

Follower plates were sometimes pushed by hydraulic fluid while the grease was being pumped out. These required excellent seals and unusual geometrical conformity, hard to obtain with commercial tanks.

Some grease trucks are essentially modified cement truck trailers with several hopper compartments. The compartments are usually

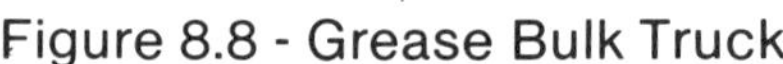

Figure 8.8 - Grease Bulk Truck

Photo courtesy of Southwest Div., Witco Corp.

insulated to slow down heat loss, since grease in a truck during a long winter haul can be quite difficult to pump off (Figure 8.8). During all seasons, bulk grease is usually shipped warm which facilitates pumping off.

One bulk grease truck consists essentially of three 3-foot (91-cm) diameter horizontal tubes, 40 feet (12.2m) long, each filled with a grease, but not necessarily all with the same grease. At one end is a piston, which is driven by air pressure so that grease is simultaneously pushed and pumped out. The piston has an effective wiping ring, leaving very little grease behind when delivery has been completed. The geometry and other details of this design appear to eliminate the problems encountered when follower plates are used.

GREASE STORAGE

Grease storage has important ramifications, depending on whether product is stored in bulk or in packages. First, some general considerations.

Grease may change its characteristics in storage. The product may oxidize, bleed, change in appearance, pick up contaminants, or become firmer or softer. The amount of change varies with length of storage, temperature, and the nature of the product.

One of the characteristics of grease, mentioned earlier, is thixotropy. This behavior--work softening followed by age hardening--varies greatly from product to product. Some high-quality products exhibit a high degree of thixotropy; on prolonged storage, they may firm (increase in consistency) more than other greases. They may also firm more in unworked than in worked consistency. Such greases are likely to slip out of consistency specification range in storage.

In addition to possible changes in consistency, greases tend to bleed--release oil--in storage. This increases with time and, generally, with temperature. Oxidation stability of modern greases is great, so prolonged storage does not lead to oxidative deterioration unless storage temperature is high.

Contaminants can deteriorate greases. Unless containers are tightly sealed, prolonged storage increases the possibility that contaminants will enter stored product. The most prevalent contaminant is moisture, which can enter apparently sealed containers of grease, rusting the containers and changing the product. Some greases, such as soda soap greases, tend to soften when contaminated with water. On the other hand, traces of moisture have been known to cause clay greases and calcium complex greases to become firmer.

Appearance changes in storage are not a common problem but can become evident if oxidation causes the surface to darken, if air slowly leaves the product improving clarity, or if water causes the grease to become hazy.

Storage in the Grease Plant

When a batch of grease has been completed and approved, large containers, such as drums, kegs, and often pails, may be filled directly from the kettle. Small containers, such as cans, tubs, and cartridges, take so long to fill that production may be slowed down. Thus the approved grease may be transferred to storage tanks, from which grease is pumped to the filling equipment and containers.

Not all grease plants can make enough grease in one batch to fill a tank truck for shipment, and scheduling two or more batches to be ready for one shipment may prove quite disruptive to manufacturing schedules, particularly when production is at a high level. Thus grease may be accumulated in storage tanks to prepare for bulk shipment.

Whether at grease plants or at customer plants, it is apparent that a large amount of grease spends some time in a storage tank. If plant storage is prolonged or temperature is kept high to ease pumping, any of the symptoms already mentioned may be experienced. Fortunately, storage time is not usually long enough to allow such deterioration to develop.

Storage in the Field

From the time that grease is packaged until it reaches the consumer, a great deal of time may elapse. Since grease can change in storage, age of the product would be a significant variable. Thus individual packages or cartons which hold small packages are usually given a date code. A common rule is that grease more than one year old should be inspected before it is sold. For unusually sensitive greases, the allowable age might be only six months. Greases known to be stable for long periods might be left unchecked for several years without much concern.

Consider the slow journey of a container of grease. It may stay in the grease plant warehouse waiting for an order and a suitable shipment. It may then go to a central warehouse awaiting distribution, then to a field warehouse awaiting a customer order. The customer then holds the material in his inventory until it is needed. Some of these warehouses are very hot; temperatures as high as 130°F (54°C) have been observed.

Product is sometimes left at the back of a warehouse, while new product is placed in front and shipped ahead of the older material. Product is sometimes stored next to a heat source, such as a steam pipe. Product is sometimes stored out of doors, where rainwater somehow enters the "sealed" package. Storing drums on their side is sometimes helpful. These are unquestionably worst-case horror stories, but they are representative of real-life situations. They indicate the need for date coding, proper storage, movement of inventory, and product checks when the grease is old.

Bulk Grease Handling

Steel mills are large consumers of lubricating greases. In modern mills, these greases, particularly multipurpose products, are handled in bulk. Greases from kettles and storage tanks in the grease plant are pumped, while still hot, into special bulk grease trucks. These are driven as scheduled to the steel mill, where storage tanks are conveniently located. Grease from the truck, still warm, is pumped into the steel plant's storage tank.

Grease is then pumped as needed through long pipelines to the working areas of the plant--hot rolling mills, cold rolling mills, slabbing mills, shear tables, etc. Here the grease travels through dispensing systems to roll necks, screw downs, conveyor bearings, etc.

The pipelines often carry grease to smaller reservoirs, from which it is dispensed to the bearings. Smaller plants or plants in which small usage operations are found may have grease delivered directly to these reservoirs. The same is true if a special grease is preferred in one area of the mill.

Many other consumers who use large amounts of grease also have it delivered in bulk. They too enjoy the savings in this mode of operation, which are found in fewer purchase orders and in minimizing drum handling, drum storage, drum return, grease waste, etc. The grease supplier establishes a lower price for grease in bulk, since he too realizes economies.

Installation cost for such a system is high--beyond the reach of most small consumers. They sometimes receive product in jumbo bins stationed closer to the area of consumption.

The connection from grease truck to storage tank should have a diameter of 3 inches (7.5 cm) or 4 inches (10 cm). Storage tanks should hold at least 20 000 pounds (9 000 kg) and are preferably set up in pairs. Grease is piped through lines of the same diameter, but to save cost, lines as narrow as 2 inches (5 cm) are sometimes used. If such lines are long, pressure drop may become excessive. Then booster pumps may be added to rebuild pressure.

Smaller lines as narrow as 1 inch (2.5 cm) in diameter have been used, especially as branch lines. When narrow lines are used, pressures build up, sometimes to excessive levels. Then softer greases of lower apparent viscosity may be required, especially at low temperature. Alternatively, pipelines may be heated using steam or electric tracers. In the long run, using adequate lines may prove less costly than the remedial actions which must be taken to decrease resulting high pressures.

Suppliers of grease can furnish apparent viscosity curves for their greases (see Chapter 3 on grease testing). These curves may be transferred to pipe flow charts available from the NLGI. From these, pressure drops through lines of different diameters can be estimated. With this information, pipe systems are designed which should give satisfactory operation when used with the grease selected.

DISPENSING AND HANDLING GREASES

Grease which has been delivered to the machine must then be applied to mechanical parts subject to friction. Such parts include bearings, gears, cams, slides, and chains. In small shops or where usage is low, lubricant may be applied by hand. Even where a bearing packer is employed, grease is often smeared by hand over the bearing, housing, and other metal surfaces before the bearing is reinstalled and the mechanism reassembled.

Roller chains and slides are sometimes lubricated with a soft grease applied with a brush, swab, pad, or roller. Another type of hand application is by grease cup, which must be filled, then turned down by hand.

Smaller volumes of grease are applied in other ways worthy of mention. Soft greases are sometimes applied by paddle or spray to slow-running open gears, wire ropes, chains, etc. Grease of normal consistency, such as NLGI No. 2, may be sprayed if enough force can be developed.

For small devices, the grease can be diluted with a solvent and the diluted product sprayed on, sometimes from an aerosol container. The solvent should be allowed to evaporate before using the machine.

Solvent-diluted grease has been used also to lubricate automotive body hardware. The parts are dipped in the grease solution, then dried, leaving a thin layer of grease. In a related manner, grease has been warmed rather than diluted, the temperature varying with the kind of grease used. Parts to be lubricated are dipped into the soft, warm grease, which penetrates to all operating points. On cooling, the parts are covered with firm grease. Some link chains have been dipped into lithium grease or similar product held at 185° to 190°F (85° to 88°C). Ball bearings have been lubricated in a vat of calcium soap grease held around 140°F (60°C).

Railroad traction motor greases are generally packaged in 1-pound (454-g) or 2-pound (908-g) polyethylene pouches, which are thrown into the gear case, bag and all. The lubricant thus consists of grease plus ground-up polyethylene.

When grease is applied by grease gun, as on an automotive chassis or small tractor, the gun may be pumped by hand or by electric motor, air pressure, etc. As machines used become larger and more complex or when a company uses a large number of machines, efficiency leads to the use of a lubrication system. These systems are designed to meet the requirements of the machine and the application; they assure that all points are lubricated and that amount of lubricant and frequency of lubrication are adequate. If investment can be warranted, sophisticated modifications are available in which cycle times and amount of lubricant dispensed can be varied and alarms--audible and/or visual--notify the user when something goes wrong.

Another consideration favoring the installation of systems is safety. Points to be lubricated are often high up, requiring the oiler to climb on the machine. They may be next to or behind a hot area or a moving part, such as an exposed gear. The machine, of course, must be shut down during such lubrication, resulting in downtime. But in the pressure to get back into production, accidents may cause bruises or burns. Lubrication systems minimize or eliminate such hazards.

Consultation among consumer, lubricant supplier, and dispensing system designer should lead to a well-lubricated, long-lasting, productive, and efficient machine. This is the payout for the effort expended by all who work in the lubricating grease industry.

GLOSSARY

Definitions of Terms Relating to the Lubricating Grease Industry

This Glossary was prepared by the Literature Subcommittee of the National Lubricating Grease Institute.

ADDITIVE - Any material added to a lubricant to impart new properties or enhance properties. Typical examples are antioxidants, corrosion inhibitors, antifoam, antiwear, and extreme pressure (EP), or antiweld additives.

ADHESION - The force or forces causing two materials, such as a lubricating grease and a metal, to stick together.

AGE HARDENING - Increase in consistency (hardening) of a lubricating grease with storage time. (See also THIXOTROPY).

ANHYDROUS - Devoid of water. With regard to a finished lubricating grease, one in which substantially no water can be detected by ASTM D 128.

ANTIOXIDANT - (oxidation inhibitor) - An additive, usually incorporated in a relatively small proportion, to retard oxidation of lubricants.

APPARENT VISCOSITY - The ratio of shear stress to rate of shear of a non-Newtonian fluid, calculated from Poiseuille's equation and measured in poises. The apparent viscosity of most greases varies with changing rates of shear and temperature, and must therefore be reported as the value at a given shear rate and temperature. This is generally measured as described in ASTM D-1092.

APPEARANCE - Those characteristics of a lubricating grease which are observable by visual inspection only. This general term includes various characteristics described under BLOOM, BULK APPEARANCE, COLOR, LUSTER and TEXTURE.

BLEEDING - the separation of liquid lubricant from a lubricating grease for any cause.

BLENDING - Blending is regarded as the process of mixing fluid lubricant components for the purpose of obtaining desired physical properties.

It is the intent to establish a difference between blending and compounding. COMPOUNDING is here regarded as the mixing or otherwise combining lubricant components with other components for the purpose of securing chemical and/or physical properties not usually obtainable by blending of fluid lubricant components alone.

BLOOM - The surface color (usually blue or green) of a lubricating oil or grease when viewed by reflected daylight at an angle of about 45 degrees from the surface. BLOOM is associated with the absorption of ultraviolet light in the oil and may not be visible if the sample is viewed by artificial light.

BOUNDARY LUBRICATION - See LUBRICATION

BULK APPEARANCE - Visual appearance of grease when the undisturbed surface is viewed in an opaque container. See also TEXTURE. BULK APPEARANCE should be described in the following terms:

BLEEDING - Showing free oil on the surface of the lubricating grease (or in the cracks of a cracked grease).

CRACKED - Showing surface cracks of appreciable magnitude. In describing such a lubricating grease, the number and size of the cracks should be included in the description.

GRAINY - A surface composed of small granules or lumps of constituent thickener particles.

ROUGH - A surface composed of many small irregularities.

SMOOTH - A surface relatively free of irregularities.

CAVITATION - Formation of a void due to reduced pressure in lubricating grease dispensing systems. CAVITATION can lead to failure of the grease to flow to the suction of the system.

CHANNELING - 1. A term used in connection with lubricating greases to describe the (usually desirable) tendency to form a channel by working down of lubricating grease in a bearing, leaving shoulders of unworked grease which serve as seal and reservoir.

CHANNELING - 2. A term used in connection with liquid lubricants and flow-type lubricating greases to describe the tendency at low temperatures, for these materials to form a plastic structure sufficiently strong to resist flow under gravitational forces only. (Similar to, but not identical with the pour point of liquid lubricants, it is measured on empirical tests such as Method 3456 in Federal Test Method Standard No. 791 (B).

COEFFICIENT OF FRICTION - The coefficient of Static Friction is the ratio of the tangential force causing initial perceptible motion in a sliding system to the load perpendicular to the motion.

The coefficient of Kinetic Friction is the ratio of the tangential force sustaining motion at constant velocity in a sliding system to the load perpendicular to the motion. The term coefficient of friction usually refers to the coefficient of Kinetic Friction.

COHESION - The molecular attraction causing particles of a given substance to attract each other and stick together. Cohesion contributes to the resistance of a lubricating grease to flow.

COLD SETTS - See SETT

COLLOID - Substances of particle size larger than molecules, but small enough to possess reasonable dispersion stability in two phase systems. A lubricating grease is a colloidal system. (See also THICKENER).

COLOR - (of lubricating grease) - The shade and intensity shown when lubricating grease is viewed under conditions such as to eliminate BLOOM.

Such conditions may be obtained by viewing the lubricating grease in an opaque container such as a metal package, by reflected light from a position approximately perpendicular to the surface; or by viewing it with transmitted light only, by placing the sample on a transparent plate. When colors of lubricating grease are referred to, the method by which colors are determined should be clearly indicated.

Colors of lubricating greases are best described in terms of the predominant hue such as amber, brown (or perhaps green, red, or blue for dyed grease) with a qualifying adjective describing intensity in terms of light, medium, or dark.

COLOR - (of lubricating oil) - That shade shown when viewed under transmitted light only. Usually lubricating oil colors are obtained by viewing the oil under specified conditions in test equipment. Several such methods are available, the most widely used being ASTM D 1500 which desribes the colors in terms of numbers.

COMPLEX SOAP - A soap wherein the soap crystal or fiber is formed usually by co-crystallization of two or more compounds:

1. The normal soap (such as metallic stearate or oleate).

2. The complexing agent. (Examples of complexing agents are the metallic salts of short chain organic acids such as acetic or lactic, or the inorganic salts such as the carbonate or chloride. The complexing agent brings about a change in grease characteristics usually recognized by an increase in dropping point).

COMPOUNDING - See BLENDING

CONSISTENCY - (HARDNESS) - The degree to which a plastic material such as lubricating grease resists deformation under the application of force. It is therefore a characteristic of plasticity, as viscosity is a characteristic of fluidity. Consistency is usually indicated by ASTM Cone Penetration, ASTM D-217 (IP 50) or ASTM D-1403.

CORROSION - The gradual destruction and/or pitting of a metal surface due to chemical attack. (See also FRETTING).

DISPENSABILITY - The property of a grease which governs the ease with which it may be transferred from its container to its point of application. Mostly used in discussion of grease dispensing systems, where it includes both the properties of pumpability and feedability.

DROPPING POINT - The temperature at which a drop of material falls from the orifice of the test apparatus under the conditions of ASTM D 566 (IP 132) and ASTM D 2265. (This test should be considered as having very limited bearing upon service performance. It is not the melting point of the grease, this being a term of little or no significance, when applied to plastic materials such as lubricating grease. These materials are characterized by a melting range during which the material becomes steadily softer. Only through the establishment of an arbitrary and fully controlled test procedure, such as ASTM D 566 (IP 132) or ASTM D 2265, can any single temperature be established as a characteristic of the material).

DRY FILM LUBRICANT - Any class of lubricants wherein the reduction of friction and wear during sliding is caused by making the shearing take place within the crystal structure of a material with low shear strength in one particular plane. Examples include graphite, molybdenum disulfide and certain soaps.

ELASTOHYDRODYNAMIC LUBRICATION - See LUBRICATION

EVAPORATION LOSS - That portion of a lubricant which volatilizes in use or in storage. Widely used test methods are ASTM D 972 and ASTM D 2595.

EXTREME PRESSURE PROPERTY - (EP) - The ability of a lubricant to reduce scuffing, scoring and seizure of contacting bearing surfaces when applied loads are high. Commonly used laboratory test measurements of EP level are Timken OK Load (ASTM D 2509 and ASTM D 2782) and Four Ball Load Wear Index (ASTM D 2596 and ASTM D 2783).

FALSE BRINELLING - localized fretting which occurs when the rolling elements of a bearing vibrate or oscillate with small amplitude while pressed against the bearing race. The mechanism proceeds in stages: 1) asperites weld, are torn apart, and form wear debris which may be subsequently oxidized; 2) due to the small amplitude the detritus cannot rapidly escape and becomes an abrasive that accelerates the wear. The resulting wear depressions appear similar to Brinell depressions obtained with static overloading. (See also FRETTING).

FEEDABILITY - The ability of a lubricating grease to flow to the suction of a dispensing pump at a rate at least equal to pump delivery capacity. (Some lubricating greases do not feed satisfactorily and cause cavitation at the inlet to a dispensing pump. In such cases, feedability can often be made satisfactory by the use of follower plates).

FIBER - (in lubricating grease) - The form in which soap thickeners occur. Some soaps crystallize in threads which are of the order of 20 or more times as long as they are thick. (Most soap fibers are microscopic in size, so that the grease appears smooth to the eye).

The greases having FIBROUS appearance are those in which the fiber bundles are large enough to be seen by the naked eye. The most common fibrous lubricating grease is sodium base, although not all sodium base greases are fibrous. (See also APPEARANCE and TEXTURE).

FIBRIL - An extremely small fiber, usually barely visible even at maximum magnification of the electron microscope. Fibrils may collect in bundles to form the larger fibers.

FILLER - A material added to a lubricant to increase bulk or density. Dependent on type and amount, a filler may contribute to, detract from, or have no effect on lubricating properties of the grease. Examples of fillers are talc, pigments, carbon black.

FILM STRENGTH - The ability of a film of lubricant to resist rupture due to load, speed and temperature.

FOLLOWER PLATE - A plate fitted to the top surface of lubricating grease in a container and so designed that as lubricating grease is dispensed, atmospheric pressure assists gravitational forces in delivering grease to the inlet of the dispensing system.

FRETTING - A form of attritive wear caused by vibratory or oscillatory motion of limited amplitude characterized by the removal of finely divided particles from the rubbing surfaces, often followed by immediate local oxidation, hence the term FRICTION OXIDATION. In the case of ferrous metals, the oxidized wear particles are abrasive iron oxide (Fe_2O_3) having the appearance of rust, which has led to the term FRETTING CORROSION. (See also FALSE BRINELLING).

FRETTING CORROSION - See FRETTING

FRICTION - The force resisting relative motion between two bodies in contact.

FRICTION OXIDATION - See FRETTING

GEL - An elastic solid mixture of a colloid and a liquid possessing a yield point and a jellylike texture.

GREASE - See LUBRICATING GREASE

HOMOGENIZATION - The process of subjecting a lubricating grease to intimate mixing and intensive shearing action, resulting in a more uniform dispersion of components.

HYDRATED SOAP - A soap which has water associated with its structure. A typical example is a water-stabilized calcium soap grease which owes its stability to hydrated calcium soap.

HYDRODYNAMIC LUBRICATION - See LUBRICATION

HYDROPHILIC - Having an affinity for water; capable of uniting with or dissolving in water.

HYDROPHOBIC - Having antagonism for water; not capable of uniting or mixing with water.

HYDROSTATIC LUBRICATION - See LUBRICATION

INCOMPATIBILITY - Two lubricating greases show incompatibility when a mixture of the products shows physical properties or service performance which are markedly inferior to those of either of the greases before mixing. Performance or properties inferior to one of the products and superior to the other may be due to simple mixing and would not be considered as evidence of incompatibility.

INDUCTION PERIOD (GREASE OXIDATION) - A period of time during which oxidation occurs at a relatively low rate. The point at which a change occurs to a significantly more rapid rate is the end of the induction period. There are several methods of measurement, including ASTM D-942.

INORGANIC THICKENER - See NON-SOAP THICKENER

INSOLUBLES - Components of a lubricating grease which are insoluble in the prescribed reagents in an analytical procedure such as ASTM D 128. The analytical procedure used should be indicated when insolubles are specified. Additional identifying analytical tests are required to determine the nature and composition of insolubles which may consist of fillers, additives, certain types of thickeners, or inadvertent impurities.

LUBRICATING GREASE - A solid to semifluid product of dispersion of a thickening agent in a liquid lubricant. Additives imparting special properties may be included.

LUBRICATING GREASE STRUCTURE - The physical arrangement of the component particles of a lubricating grease thickener, additive (if any) and liquid lubricant. It is the nature and stability of this arrangement which determine the appearance, texture and physical properties of lubricating grease.

LUBRICATION - The reduction of friction or wear between two load-bearing surfaces, in relative motion, by the application of a lubricant. Four types of lubrication are described below.

BOUNDARY LUBRICATION - The condition in which the lubricant film becomes too thin to give full fluid film separation of the rubbing surfaces, and, as a result, surface asperities come in contact. Consequently, friction and wear protection properties are determined by the chemical nature of the lubricant rather than its bulk properties.

ELASTOHYDRODYNAMIC LUBRICATION - The condition in which surfaces of heavily loaded machine elements are either completely or in part separated by a very thin lubricant film. The elastic deformation of the contacting surfaces traps the lubricant subjecting it to high pressure which increases its viscosity and its load carrying capacity.

HYDRODYNAMIC LUBRICATION - The condition in which the shape and relative motion of the sliding surfaces cause the formation of a continuous fluid film under sufficient pressure to prevent any contact between the surfaces. It is commonly called Fluid Film Lubrication.

HYDROSTATIC LUBRICATION - That state of lubrication in which the lubricant is supplied to a plain bearing under sufficient external pressure to separate the opposing surfaces by a continuous lubricant film.

LUSTER - The intensity of light reflected by lubricating grease; its sheen or brilliance. LUSTER should be described as follows:

BRIGHT - Reflects light with a relatively strong intensity.

DULL - Reflects light with a relatively weak intensity. Some greases with a high water content may have a dull luster. Certain thickeners and fillers give a grease a characteristic dull luster.

MECHANICAL STABILITY - See SHEAR STABILITY

NEWTONIAN BEHAVIOR - The property of a simple liquid in which the shear rate is directly proportional to the shear stress. This constant proportion is the viscosity of the liquid.

NLGI NUMBER - A numerical scale for classifying the consistency range of lubricating greases, and based on the ASTM D 217 worked penetration at 25°C (77°F). NLGI NUMBERS are in order of increasing consistency (hardness) as follows:

NLGI NUMBER	ASTM WORKED PENETRATION
000	445-475
00	400-430
0	355-385
1	310-340
2	265-295
3	220-250
4	175-205
5	130-160
6	85-115

(Greases both softer and harder than these consistency ranges and numbers are well known in the industry. Such greases do not bear an NLGI Number).

NON-NEWTONIAN BEHAVIOR - The property possessed by some fluids and many plastic solids, including lubricating grease, of having a variable relationship between shear stress and rate of shear. (NON-NEWTONIAN materials, therefore, do not possess a viscosity as defined by Newton, but rather an apparent viscosity, the quantitative value of which may vary widely with varying shear rate. Conventional types of viscometers with uncontrolled shear rates will not satisfactorily measure NON-NEWTONIAN materials).

NON-SOAP THICKENER - Any of several specially treated or synthetic materials, excepting the metallic soaps which can be either thermally or mechanically dispersed in liquid lubricants to form lubricating grease. Sometimes called SYNTHETIC THICKENER, INORGANIC THICKENER, OR ORGANIC THICKENER.

OILINESS AGENT - A material which reduces friction by formation of an adsorbed film.

ORGANIC THICKENER - See NON-SOAP THICKENER

OXIDATION STABILITY - The resistance of lubricants to chemical reaction with oxygen. The absorption and reaction of oxygen may lead to deterioration of lubricants. Several testing methods are in use, including ASTM D 942 (IP 142).

PENETRATION - An arbitrary measure of consistency (hardness), based on ASTM D 217 (IP 50) (and similar methods standardized by these and other organizations). All penetration measurements are in an inverse scale of consistency, that is, the softer the consistency, the higher the penetration number.

The ASTM Definitions are given as follows:

PENETRATION of lubricating grease is the depth in tenths of a millimeter, that the standard cone penetrates the sample under prescribed conditions of weight, time and temperature.

UNDISTURBED PENETRATION is defined as the penetration at 25°C (77°F) of a sample of grease in its container with no disturbance.

UNWORKED PENETRATION is the penetration at 25°C (77°F) of a sample of lubricating grease which has received only minimum disturbance in transferring to a grease worker cup or dimensionally equivalent container.

WORKED PENETRATION is the penetration of a sample of lubricating grease which has been brought to 25°C (77°F), subjected to 60 double strokes in a standard grease worker, and penetrated without delay.

PROLONGED WORKED PENETRATION is the penetration of a sample of lubricating grease after it has been worked more than 60 double strokes in a standard grease worker at a temperature of 15° to 30°C (59° to 86°F). After the prescribed number of double strokes, the worker and contents are brought to 25°C (77°F), worked an additional 60 double strokes, and penetrated without delay.

BLOCK PENETRATION is the penetration at 25°C (77°F) of a sample of lubricating grease which is sufficiently hard to hold its shape, determined on the freshly prepared face of a cube cut from a block of the grease.

PLASTICITY - That property of apparently solid material which enables it to be permanently deformed under the application of force, without rupture. (Plastic flow differs from fluid flow in that the shearing stress must exceed a yield point before any flow occurs).

PUMPABILITY - The ability of a lubricating grease to flow under pressure through the line(s), nozzle(s), and fitting(s) of grease dispensing system(s). It is best indicated by the apparent viscosity at moderate rate of shear. See ASTM D 1092 and NLGI Publication, NLGI Steady Flow Charts for Grease.

REVERSIBILITY - The ability of a grease to return to its normal grease-like consistency after temporary exposure to temperatures near or above the dropping point of that grease. Only a few types of greases have this property.

RHEOLOGY - Study of the deformation and/or flow of matter in terms of stress, strain, temperature and time. The rheological properties of lubricating greases are commonly measured by PENETRATION and APPARENT VISCOSITY.

RHEOPECTIC GREASE - A lubricating grease which has the property of increasing in consistency, (hardening appreciably), upon being subjected to shear.

SAPONIFICATION - The interaction of fatty acids, fats or esters generally with an alkali to form the metallic salt. This salt is commonly called soap.

SET - In the manufacture of a lubricating grease, the change from a fluid to a semifluid or plastic state.

SETT - The word "Sett" is uniquely applied to a particular type of product, i.e., the Cold Sett greases which change from a fluid to a semifluid or plastic state after component combination and often after packaging.

SHEARING - Slipping or sliding of one part of a substance relative to an adjacent part. In a solid, such action involves cutting or breaking of the crystal structure, but in a fluid or plastic, shearing does not necessarily destroy the continuous nature of the substance.

SHEAR RATE - The rate of a slip within a substance engaging in flow. The average or mean shear rate in a pipe or tube is proportional to the average velocity divided by the radius of the tube. It, therefore, has the dimensions of the reciprocal of the time and is usually expressed in the unit of reciprocal seconds (sec^{-1}). The mean shear rate is reported in the determination of apparent viscosity in ASTM D-1092.

SHEAR STABILITY - The ability of a lubricating grease to resist changes in consistency (hardness) during mechanical working. Working may be in any of several types of laboratory machines or may be in actual service. This may also be called MECHANICAL STABILITY.

SHEAR STRESS - The force per unit area tending to cause shearing in a substance. In fluids, the ratio of the shear stress to the shear rate is the viscosity of the substance.

SLUMPABILITY - See FEEDABILITY

SOAP - See THICKENER; COMPLEX SOAP; SAPONIFICATION.

SOL - A suspension of particles of colloidal dimensions in a liquid. These systems possess the gross properties of a liquid.

SQUEEZE FILM LUBRICATION - That state of lubrication in which surfaces thickly coated or flooded with lubricant move toward each other at sufficient speed to develop fluid pressure sufficient to support a load of short duration. Because of viscosity (or apparent viscosity), the lubricant cannot immediately flow away from the area of contact. This action occurs, for example, between gear teeth and between wrist pins and their bushings.

SYNERESIS - Loss of liquid lubricant from a lubricating grease due to shrinkage or rearrangement of the structure. The shrinkage may be due to either physical or chemical changes in the thickener. SYNERESIS is a form of BLEEDING.

SYNTHETIC GREASE - A grease composition in which the liquid lubricant is other than mineral oil.

SYNTHETIC THICKENER - See NON-SOAP THICKENER

TEXTURE - That property of lubricating grease which is observed when a small separate portion of it is pressed together and then slowly drawn apart. TEXTURE should be described in the following terms:

BRITTLE - Has a tendency to rupture or crumble when compressed.

BUTTERY - Separates in short peaks with no visible fibers.

LONG FIBERS - Shows tendency to stretch or string out into a single bundle of fibers.

RESILIENT - Capable of withstanding moderate compression without permanent deformation or rupture.

SHORT FIBER - Shows short break-off with evidence of fibers.

STRINGY - Shows tendency to stretch or string out into long fine threads, but with no visible evidence of fiber structure.

(Other terms, such as SMOOTH, ROUGH, GRAINY, etc., are defined under BULK APPEARANCE).

THICKENER - The solid particles which are relatively uniformly dispersed to form the structure of lubricating grease in which the liquid is held by surface tension and other physical forces. (The solid particles may be fibers, as is the case with various metallic soaps, or plates or spheres, as is the case with some of the non-soap thickeners. The only general requirements are that the particles should be extremely small and that they be capable of uniform dispersion in the lubricants.)

THIXOTROPY - (in lubricating grease) - That property which is manifested by a decrease in consistency, or softening, as a result of shearing, followed by an increase in consistency, or hardening, beginning after shearing is stopped. (Thixotropic age hardening is a relatively prolonged process proportional to aging time and is seldom, if ever, complete; whereas the apparent viscosity increase which occurs in non-Newtonian systems with decreasing shear rate is instantaneous and fully reversible. Lubricating grease is both thixotropic and non-Newtonian.)

WATER RESISTANCE - The ability of a lubricating grease to withstand the addition of water to the lubricant system without adverse effects. WATER RESISTANCE is generally considered to be made up of four components as listed below:

WASHOUT RESISTANCE - The ability of a lubricating grease to resist being removed from a bearing when operated fully or partially submerged in water. (Generally measured by ASTM D 1264).

WATER ABSORPTION CHARACTERISTIC - The characteristics of a lubricating grease when water is added to the lubricating system. WATER ABSORPTION CHARACTERISTICS may be measured by any of several suitable tests in which the lubricating grease may react in any of three ways, described as follows:

WATER SOLUBLE - The lubricating grease absorbs the water, and then de-gels to a semifluid consistency.

WATER ABSORBENT - The lubricating grease absorbs relatively large quantities of water with small or no change in consistency, and without leaving free water as a separate phase.

WATER RESISTANT - The lubricating grease does not absorb

more than small amounts of water, does not change appreciably in consistency, and leaves the added water as a second phase in the system.

WATER CORROSION RESISTANCE - The ability of a lubricating grease to prevent corrosion of surfaces when water is present in the lubricating system. May be measured either statically by any of a number of standard tests, or dynamically by actual operation of bearings with water added to the lubricant reservoir. (Refer to ASTM D 1743).

WATER SPRAY RESISTANCE - The ability of a grease to resist displacement from a surface by the impact of water spray. The method of test used to evaluate this characteristic for lubricating greases is given in ASTM D 4049.

Lubricating greases for various types of service may need any of the several types of WATER RESISTANCE characteristics described above, so that they are not measures of quality except for specific situations where particular properties are required.

WEAR - The removal of materials from surfaces in relative motion. Three types of wear are described below:

ABRASIVE WEAR - Removal of materials from surfaces in relative motion by a cutting or abrasive action of a hard particle (usually a contaminant).

ADHESIVE WEAR - Removal of materials from surfaces in relative motion as a result of surface contact. Galling and scuffing are extreme cases.

CORROSIVE WEAR - Removal of materials by chemical action.

WORKING - The subjection of lubricating grease to any form of agitation or shearing action beyond simple transfer to any test apparatus.

YIELD - (of lubricating grease) - The amount of grease of a given consistency which may be made with a definite amount of thickening agent. As the yield increases, percent thickener decreases.

YIELD POINT (or YIELD VALUE or YIELD STRESS) - The minimum shear stress required to produce flow of a plastic material. It can be estimated by the intercept on the shear stress axis of the shear stress - shear rate curve, by extrapolation of the straight portion of the curve.

RELATED REFERENCES

A complete compilation of references seems neither necessary nor desirable. Several reference sources of particular interest and merit are given here.

E. N. Klemgard. *Lubricating Greases: Their Manufacture and Use.* New York: Reinhold Publishing Corporation, 1937.

An encyclopedic work. General information still pertinent. Most of the specific formulations given are now obsolete.

C. J. Boner. *Manufacture and Application of Lubricating Greases.* New York: Reinhold Publishing Corporation, 1954.

Another encyclopedic work. Again, much that is pertinent, much that is obsolete. Invaluable for reference.

C. J. Boner. *Modern Lubricating Greases.* Scientific Publications (GB) Ltd. Not dated. Preface dated 1976.

Not as complete as the earlier work, but effectively points out changes in the industry in the intervening 20 years.

R. S. Barnett. "Review of USA Publications on Lubricating Grease." NLGI *Spokesman*, 35, No. 6, 180-187. September 1971.

E. R. Booser, Editor. *Handbook of Lubrication.* CRC Press, Inc. (two volumes) 1983.

This book discusses many areas of lubrication. Of necessity, coverage of lubricating greases is limited.

Annual Book of ASTM Standards, Volumes 05.01, 05.02, and 05.03 (revised annually). These volumes cover Petroleum Products and Lubricants.

W. P. Scott. "Index of the NLGI *Spokesman,* April 1950-May 1981."

Invaluable index to all the articles published in the *Spokesman* in over 30 years. Available from NLGI International Office.

NLGI PUBLICATIONS

Recommended Practices

Lubricating Passenger Car Wheel Bearings
Lubricating Truck Wheel Bearings
Passenger Car Ball Joint Front Suspension
Line Strainers for Centralized Dispensing Systems
Standard Unit of Measurement for...Centralized...Systems
Viscosity of the Base Fluid in Lubricating Grease
Storage of Lubricating Grease

Other NLGI Literature

Chassis and Wheel Bearing Service Classification System
Classification of Lubricating Grease
Steady Flow Charts for Grease
Glossary of Terms
Grease Production Survey Reports
NLGI Information Brochure

INDEX

Glossary references are indicated by italics.

NOTES

NOTES

NOTES

NOTES